Citrus Growing

in

FLORIDA

Looking north from entrance to 2,000-acre planting, 10 years of age, of East Coast Division, Lake Hamilton Citrus, Inc. (located in St. Lucie County).

Courtesy Carl B. Christopher

Citrus Growing in FLORIDA

Revised Edition

LOUIS W. ZIEGLER and HERBERT S. WOLFE

A University of Florida Book

The University Presses of Florida
Gainesville

Library of Congress Cataloging in Publication Data

Ziegler, Louis W., 1907–
Citrus growing in Florida.

"A University of Florida book."
Bibliography: p.
Includes index.
1. Citrus fruits—Florida I. Wolfe, Herbert Snow, joint author. II. Title.
SB369.2.F6Z53 1975 634'.3 75-5664
ISBN 0-8130-0488-8

Second Printing, 1979

PRINTED BY STORTER PRINTING COMPANY
GAINESVILLE, FLORIDA

Table of Contents

Preface	vii
Preface to the First Edition	ix
1. *Origin and History*	1
2. *The Kinds of Citrus Fruits*	12
3. *Propagation of Citrus Fruits*	63
4. *Climate and Soil in Citrus Growing*	83
5. *Planting Citrus Trees*	111
6. *Bringing Citrus Trees into Production*	126
7. *The Bearing Grove and Efficient Performance*	136
8. *Fertilizing Bearing Citrus Trees*	152
9. *The Citrus Spray Program*	172
10. *Other Grove Maintenance Practices*	193
11. *The Crop—Harvesting, Maturity, and Grade*	217
12. *Citrus Fruits for the Home*	232
Appendix	239
Index	241

Preface

THE citrus industry of Florida has undergone tremendous changes during the past fourteen years. Continued urbanization of the state has had a great influence on land values, taxes, and labor availability. Citrus plantings have been pushed out of some of the historically highly desirable locations by the expansions of urban and industrial developments; these plantings have been replaced with grove developments on flatwood and swamp soils in areas not previously occupied by the industry. Approximately one-half of the present 877,000 acres of citrus require drainage to some degree. Problems caused by viruses and nematodes have been more distinctly identified and described; psorosis is not the serious problem of former years yet tristeza still poses a considerable threat. Purple scale and Florida red scale are now of minor importance whereas snow scale has invaded practically all citrus areas. Young tree decline (YTD), cause yet unknown, has become the newest scourge affecting the industry.

Much new knowledge of citrus has been obtained through the sincere efforts of researchers and growers. My gratitude is expressed to all workers who have contributed to the vast literature. Particular thanks must be extended to F. P. Lawrence and Dr. L. K. Jackson of the Florida Cooperative Extension Service, G. D. Bridges of the Citrus Budwood Registration Bureau, Dr. J. H. O'Bannon of the U.S. Department of Agriculture, Drs. A. H. Krezdorn (Chairman) and J. Soule of the Fruit Crops Department of the University of Florida Institute of Food and Agricultural Sciences, and my son, M. R. Ziegler, who offered many helpful suggestions. The author of this revised edition assumes full responsibility, however, for the views herein expressed.

LOUIS W. ZIEGLER

March 1975

Preface to the First Edition

WITH the exception of the tourist business, citrus production is the leading industry of Florida. In addition to the large-scale individual growers, cooperatives, caretaking organizations, and corporations that play the major role in this industry, there are a great many growers who either manage small acreages themselves or have groves under professional management. Furthermore, thousands of home gardeners cultivate citrus trees for a home supply of fruit as well as for ornament.

The authors have had especially in mind these last two groups, and have written this book largely for their benefit, in the hope that it may make it easier for them to understand the scientific basis of our modern citriculture. In the past half-century, fruit growing has changed steadily from an activity in which the amateur could compete easily with the professional to one in which only those who were well-grounded in horticultural science and its applications to each fruit could hope to succeed. The present book endeavors to set forth those principles and their applications for citrus growing in a comprehensive but easily understood manner. The authors earnestly hope that it will be read by the professional citriculturist with approval, and that the small grower and homeowner will find it a very real help and guide to modern citrus production in Florida.

Although the authors have each had over thirty years of firsthand experience with citrus, they recognize freely that much of the information contained in this book has come from the published and unpublished work of others.

To their colleagues in the Citrus Experiment Station of the University of Florida and in the Horticulture Field Station of the United States Department of Agriculture, the authors acknowledge deep indebtedness for much of what is known about the science of citrus production. Many growers, too numerous to mention, have made valuable contributions over the years. Various production managers and nurserymen have given freely of information and other helps. To all of these, grateful appreciation is rendered. A special debt of gratitude is owed to Mr. R. E. Norris, Lake County Agricultural Agent, and Mr. F. P. Lawrence, Citriculturist of the Florida Agricultural Extension Service, who were kind enough to read the manuscript critically and make helpful suggestions for its improvement. The authors assume full responsibility, however, for the views herein expressed.

LOUIS W. ZIEGLER
HERBERT S. WOLFE

1. Origin and History

Origins and Introduction in the Old World

THE United States leads all other countries in citrus production, and Florida leads all other states in such production. Indeed, Florida alone produces more citrus fruits than any country outside the borders of the United States. Furthermore, citrus fruits of some kind can be grown in every county in this state, although the choice of kinds is rather limited in some areas. There is, therefore, a keener interest in citrus fruits—whether for commercial production or only to supply the home—in Florida than in any other state.

The original home of nearly all species of citrus was probably on the warm southern slopes of the Himalayas in northeastern India and adjacent Burma. Long before the dawn of history, primitive forms of the orange, shaddock, and some mandarins crossed over the Himalayan passes into western China—whether by the activity of primitive man or by other means, we will never know—and developed secondary centers of distribution in southern China and Indo-China. Lime, lemon, and citron remained south of the mountains, as did some of the mandarins, and spread south into India and eastward into the Malayan region. Only the trifoliate-orange (*Poncirus*) and the kumquat (*Fortunella*) among citrus fruits seem to have originated in China, and their starting place was in the eastern part of that country.

The sweet orange must have developed its present characters in southwestern China. Like the shaddock it is not known anywhere in the truly wild state, and quite possibly it originated from other citrus species under cultivation and has always been a garden plant. It is impossible to say how long it has been cultivated in China because written records do not go back very far in that country. The earliest mention of oranges occurs in a book compiled in the sixth century B.C., the *Shu-ching*, which purports to be taken from ancient records going back to before 2000 B.C. Chinese historians have long been aware, however, that any records dated earlier than 1000 B.C. (except inscriptions on bones used in divination) are highly suspect, and that this particular reference cannot be credited to a period earlier than 600 B.C.

In the *Chou-li*, dealing with the Chou regime around 700 B.C. but actually compiled some four hundred years later, there is a distinction made between sweet and sour oranges, whereas the previous reference spoke only of oranges. There is some other reason to think that the sour orange only reached the Yangtze Valley around 400 B.C., whereas sweet oranges may have been cultivated from much earlier times. Mandarins, as distinct from oranges, are first mentioned in Chinese literature about 200 B.C. The shaddock was included in the earliest mention of oranges, but citron was apparently not known until the fourth century A.D. and lemons not until the tenth century.

India seems to be the region where lemon, citron, and some species of mandarins have developed their present form. Sanskrit literature of the eighth century B.C. mentions both citron and lemon, whereas no Sanskrit name is known for the shaddock, and that of the orange dates only from the first century A.D.; however, we shall see that oranges had reached India long before this. This Sanskrit name for the orange, *nagarunga*, became *naranj* in Persian, *aurantium* in Latin, *naranja* in Spanish, and *orange* in English.

The citron and lemon (of some type) have greater written evidence of antiquity than the orange, although they are not necessarily older in cultivation. The evidence for reference to citron in Egypt in 1500 B.C. is very dubious. The lime is known truly wild only in the Malay peninsula, from which origin it apparently spread westward into India and eastward into the Pacific Islands.

The first citrus fruit known to Europe was undoubtedly the citron, prized for its fragrant rind. The naturalists who accompanied Alexander on his conquest of Persia and northern India brought back descriptions of the tree and its culture which were made widely known by Theophrastus, around 300 B.C., but this was not the first acquaintance of the Greeks with the fruit. At least fifty years before this, citrons were being imported into Athens from Persia, and since the Greeks called it first the Median-apple, it seems probable that they had known it when the Medes ruled Persia, before the sixth century B.C. Greek colonists apparently introduced its culture into Palestine about 200 B.C. Eventually the citron replaced the cedar cone as the "fruit of the goodly tree" authorized for use in the Jewish Feast of Tabernacles, and Tolkowsky advances cogent arguments for this transition having been ordered by Simon the Maccabee in 136 B.C. The significance of this event for us lies in the probable explanation of the origin of the word *citrus*. The Greek word *kedros* originally meant the cedar, but apparently as the result of the substitution of citrons for cedar cones in this ceremony, the Palestinian Greeks began to say cedar-apple (*kedromelon*) instead of Median- or Persian-apple. The Greek *kedros* became Latinized as *cedrus*, and this changed to *citrus*. By the time of Pliny (A.D. 70) *citrus* meant the citron tree. Of course, our word *citron* is derived from *citrus*, not the other way around, and *citreum*, *malum citreum*, or *malum medicum* were Pliny's names for the fruit. Cultivated in Asia Minor and Italy in the first century A.D., the citron was being grown in Greece a century later, and it seems to have been widely used in Moslem medicine in Egypt and Spain before A.D. 900, perhaps very long before.

The sweet orange in its present form must have reached India from China well before the beginning of the Christian era, for in the first century A.D., sweet oranges were known to the Romans as Indian fruit. Certainly by the end of that century oranges were being imported into Rome, probably from Palestine and Egypt, where plantings had been established by seeds from India; for the time required for the journey from India to Italy was much too long for oranges to survive it. There is evidence that orange trees grew in southern Italy at that time, but apparently they could not mature fruit. By 300, oranges were being picked there, however; but in the Dark Ages following the sack of Rome by the Goths and Vandals, they may have been lost.

In Moorish gardens of southern Spain around the year 900, oranges were quite common, but it is not possible to be sure that some of these were sweet oranges, for accounts of citrus fruits rarely distinguished between sweet and sour forms. However, sweet oranges were certainly abundantly available in Bagdad and Cairo, and the Spanish Moors imported a great variety of exotic plants from Iraq and Egypt for the adornment of their gardens. It hardly seems possible that they did not bring in sweet as well as sour oranges.

Sweet oranges were undoubtedly cultivated in southern Europe long before Vasco da Gama reached India in 1498. The most direct evidence is a letter written in 1483 in which Louis XI of France asked that sweet oranges be sent him from Provence; but there is plenty of indirect evidence. It is well known that Columbus took orange, lemon, and citron seeds from the Canary Islands with him to Hispaniola on his second voyage in 1493, and while we have no statement as to whether the oranges were sweet or sour, we do have Oviedo's testimony that barely thirty years later sweet orange trees were abundant in Hispaniola. Furthermore, on the return of da Gama from India, his men reported that the sweet oranges they found were superior to those they had at home. These superior types of sweet orange, when brought back to Portugal, made cultivation of this fruit much more worthwhile than it had previously been considered. Henceforth the sweet orange took the lead, and with the development of special houses* (later called "orangeries") for growing them where cold made their outdoor culture impossible, growing of sweet oranges was extended into northern Europe in the seventeenth century. It may be noted, incidentally, that when the Portuguese finally introduced oranges directly from China to Europe in 1640, these were again so much better in quality than the type brought from India that "China" oranges soon replaced "Portugals" as the preferred type.

Introduction to and Development in Florida

The introduction of sweet oranges to Florida probably coincided with the establishment of the colony at St. Augustine in 1565, since we know of no earlier permanent settlement anywhere in the state. By 1579 Menéndez could report that oranges were becoming abundant there. Thus our Florida citrus industry is four centuries old. From the trees cultivated by the Spaniards, Indians carried fruit to their villages and scattered the seeds widely as they ate the oranges. In this way arose the many scattered thickets of wild orange trees of which remnants—all of the hardier sour type—may still be found around the lakes and along the rivers in some sections of the state. Urbanization, however, has destroyed many of these thickets. The orange proved so well adapted and flourished so well as a naturalized tree, escaped from cultivation, that William Bartram in his exploration of northern Florida in 1773–74 found great areas covered by apparently wild trees. His uncertainty as to whether they were wild or escaped is shown when he wrote: "Whether the orange tree is an exotick, brot in here by the Spaniards, or a Native to this country, is a question. I have inqu[ired] of some of the old Spaniards at Augustine, who tell me that they were first brot in by the Spaniards and spread over the country by the

*There is evidence of such culture in Italy even in the first century for citrons, using mica glazing; and in the fourteenth century these became popular in northern Italy.

Ind[ians]." It is only in relatively recent times, however, that the citrus fruits have taken a prominent place in our diet, not only as valuable sources of health-giving vitamins but also as welcome and useful additions to the pleasure of eating.

Plantings of the sixteenth and seventeenth centuries were limited to the few areas along the coast where Spanish colonies were established, and were not at all commercial. Not until Spain ceded Florida to England in 1763 was there any commerce in oranges, but under English rule and with English colonies to trade with, export of sour oranges assumed some importance to St. Augustine and orange plantings were made in several areas of East Florida. With the cession of Florida to the United States in 1821, the development of sweet orange

FIG. 1. This canopied seedling orange grove was 83 years old when photographed in 1960. It had been frozen to the ground in the freeze of 1895. Recently, it has been hedged and topped and is still producing fruit.

Courtesy G. O. Nordmann

groves moved rapidly ahead, especially along the St. Johns and its tributaries, which provided transportation to markets in the North. Introduction of long scale in 1838 and its subsequent spread to all citrus groves in the state caused a serious decline in citrus growing between 1840 and 1870, when the pest had ceased to be a serious problem.

The first big expansion came in the 1870's, as growers realized the size of the potential market and the possibility of satisfying it with Florida fruit. In the years 1874 to 1877, about 200 million oranges with a value of over $2 million were imported annually into the eastern United States, mostly from the Mediterranean area but partly from the West Indies. Imports of lemons were almost as large. All of these could have been grown in Florida, the leading growers urged, and so plantings were made apace to achieve this end. In this period there was extensive topworking of wild sour orange trees in the hammocks of north-central Florida, and here was the center of the citrus industry. The "big freeze" of 1894–95, however, caused this center to move south a hundred miles or so.

The change from growing of citrus fruit on a small scale, often as a crop of secondary interest to the grower, to its production on the large commercial scale now characteristic of Florida has occurred during the period since 1900. Several factors have influenced this phenomenal change. There is a very large area with satisfactory climatic and soil conditions for successful production of citrus fruits. Populous areas are relatively close to Florida, making for easy marketing of the crop. The development of three major rail systems out of the many short lines built during the nineteenth century has provided much more efficient transportation for the crop than the earlier water routes afforded. And more recently the availability of good roads everywhere has enabled transportation by truck to open new markets and reduce hauling costs to many points. Due tribute must also be paid to the enterprise of those growers of the last century who pioneered in the change from seedlings to budded varieties, and to the acumen shown in some of the seedling selections which have become important segments of our industry. And, finally, the fortunate choice of adopting 'Rough' lemon for use as a stock must be credited with much of the amazing citrus development of Florida, for without this stock the rapid expansion of the industry in the sandy hills of central Florida could hardly have come to pass.

Change is the character of human existence. Florida, during the decades of the 1950's and 60's and continuing to the present, has become the mecca of large influxes of new residents to which have been added the impacts of moon explorations from Cape Canaveral and the development of Disney World. Vast acreages have been developed to accommodate people, especially in the better-drained areas of the state. Land values and taxes have increased to the ultimate exclusion of many acres of groves. Citrus plantings were expanded to the poorly drained areas until a total of almost 1,000,000 acres of groves existed in the late 1960's. However, in the 1970's the number of acres being planted has not kept pace with removals required by urbanites so that the total as of 1973 was an estimated 877,000 acres. As Florida lost its distinctive agricultural status of earlier years due to increasing population, the citrus industry was faced with problems of labor shortages and restrictions of road, land, and chemical usages and has been

TABLE 1
CITRUS PRODUCTION IN FLORIDA
(in 1,000 1¾ bushel boxes)

Kinds of Citrus Fruit	1968–69	1969–70	1970–71	1971–72	1972–73
Oranges, Early and Midseason	69,700	72,900	82,100	68,800	90,000
Oranges, Late	60,000	64,800	60,200	68,200	80,000
Grapefruit, Seedy	12,200	9,500	11,800	10,900	10,200
Grapefruit, White Seedless	17,000	17,700	20,200	23,800	23,600
Grapefruit, Pink Seedless	10,700	10,200	10,900	12,300	11,900
Tangerines	3,400	3,000	3,700	3,200	3,000
Temples	4,500	5,200	5,000	5,300	5,100
Tangelos	1,800	2,500	2,700	3,900	3,500
Murcotts	1,100	980	970	1,600	900
Limes	700	725	880	1,100	1,100
Total	181,100	187,505	198,450	199,100	229,300

Source: Florida Crop & Livestock Reporting Service, Orlando, Florida

relegated to a secondary position in many counties such as Orange and Pinellas.

Successful development of a fruit industry, however, requires more than a tasty product, favorable growing conditions, and facilities for marketing it. It must be possible for the grower to offer a product which will consistently merit favor by the purchasing public, to produce it at a profit, and to distribute it to a wide market in orderly fashion. Research by state and federal agencies, as well as by private organizations and individuals, has been an important element in the growth of a sound industry by overcoming many production and marketing difficulties. And the various inspection agencies, regulatory

Table 2

UNITED STATES CITRUS PRODUCTION
(in 1,000 boxes)

State	1967–68	1968–69	1969–70	1970–71	1971–72
			Oranges		
Florida	100,500	129,700	137,700	142,300	137,000
California	19,150	44,300	39,000	37,500	43,300
Texas	1,800	4,500	4,200	6,200	5,800
Arizona	3,120	5,380	4,630	3,560	4,900
Total Oranges	124,570	183,880	185,530	189,560	191,000
			Grapefruit		
Florida	32,900	39,900	37,400	42,900	47,000
California	4,618	5,060	5,250	5,040	5,100
Texas	2,800	6,700	8,100	10,100	9,200
Arizona	3,740	2,510	3,160	2,520	2,540
Total Grapefruit	44,058	54,170	53,910	60,560	63,840
			Tangerines		
Florida*	9,550	10,800	11,680	12,370	14,000
California	560	640	760	1,140	600
Arizona	150	170	350	390	570
Total Tangerines*	10,260	11,610	12,790	13,900	15,170
			Lemons		
California	13,600	12,300	12,300	13,300	13,600
Arizona	3,250	3,510	2,820	3,150	3,080
Florida	610	970	1,510	1,200	1,380
Total Lemons	17,460	16,780	16,630	17,650	18,060
			Limes		
Florida	720	700	725	880	1,100
Total all Citrus	197,068	267,140	269,585	282,550	289,170

Source: Florida Agricultural Statistics, Citrus Summary, 1972

*Including Temples, tangelos, and Murcotts in Florida since these tangerine hybrids are being produced in increasing amounts, as shown in Table 1, to supply the market for specialty citrus fruits, a market formerly supplied, almost exclusively, by the 'Dancy' tangerine.

bodies, and trade organizations have helped assure a more orderly flow of high-quality fruit to market.

Modern Florida citrus production, distinct and vastly different from that of the nineteenth century, is the result of the influence of all these —at times totally unrelated—factors. Today citrus fruits are no longer a luxury; they are essential components of the modern diet all over this country.

Citrus Growing in Other Parts of the World

In the United States, the citrus industry of California is second only to that of Florida, with Texas in third place and Arizona in fourth. California produces the bulk of the lemon crop of this country, has a very important production of sweet oranges, a small grapefruit crop, and a very small production of mandarins and limes. Texas is the second state in grapefruit production, and has a fair crop of oranges and lemons. Arizona produces lemons and grapefruit chiefly, but is developing orange plantings. There are some oranges grown commercially in Louisiana.

Mexico has considerable production of oranges and leads the world in limes. Cuba and Puerto Rico have important commercial crops of oranges and grapefruit, and all the other islands of the West Indies grow oranges and limes in abundance for their own or for minor export. The same may be said of the Central American countries.

Although Japan is second in world production of sweet oranges and mandarins combined, the latter far exceed the former. Spain is the second country in sweet orange production and has commercial crops of lemons and sour oranges. Italy leads the world in lemon production, with the United States second, and also has large crops of oranges. Greece is the only other European country with a citrus industry, producing both oranges and lemons.

Brazil leads the other South American countries and is third in world production of oranges. Argentina has important crops of oranges and lemons. Paraguay, Bolivia, and Peru grow oranges on a large scale, while Uruguay and Chile produce small amounts of grapefruit as well. Every country in South America has good supplies of citrus fruits for home consumption.

Among Asiatic countries, China may have the largest production of sweet oranges, but mandarins are grown in far larger amounts there. Shaddocks and lemons are also produced in large amounts. India has considerable commercial production of citrus fruits, and an even larger production by small growers. The mandarins are more popular there than oranges, and both limes and lemons are also very widely grown. Japan has a very important mandarin industry, leading the world in satsuma production, and a much smaller crop of oranges. Throughout the Asiatic islands, from Taiwan and the Philippines to Sri Lanka, and in the countries from Vietnam to Burma, citrus fruits

TABLE 3

RECORDED WORLD PRODUCTION OF CITRUS FRUITS, 1970*
(in 1,000 boxes)

Country	Oranges and Mandarins	Grapefruit	Lemons	Total
United States	248,832	61,932	16,450	327,214
Japan	94,492			94,492
Spain	69,435	190	2,600	72,225
Brazil	65,130			65,130
Italy	50,400		22,324	72,724
Argentina	39,528	3,960	5,761	49,249
Israel	34,400	10,040	1,200	45,640
Mexico	32,124	524		32,648
Morocco	23,709	148	88	23,945
Turkey	15,747		3,700	19,447
South Africa	15,136	2,075	382	17,593
Algeria	14,000	125	400	14,525
Greece	13,300		3,900	17,200
Australia	9,322	341	947	10,610
Lebanon	6,393		1,944	8,337
Cyprus	5,300	1,680	900	7,880
China (Taiwan)	4,913			4,913
Jamaica	2,751	635		3,386
Tunisia	2,000		232	2,232
Chile	1,354		1,160	2,514
British Honduras	1,029	380		1,409
Surinam	378	163		541
Trinidad and Tobago	343	420		763
New Zealand		85	86	171
World Total	750,016	82,698	62,074	894,788

Source: *Agricultural Statistics*, 1972, U.S. Department of Agriculture.

*Citrus fruits are grown extensively in India, mainland China, and other Asian countries; limes are grown in commercial quantities, in descending order, in Mexico, Egypt, and the United States with some production in most tropical countries. Unfortunately, current production data are unavailable.

play an important part in the food supply of the people, but production is not found on a large scale.

In the Near East, Israel has the largest citrus industry, producing both oranges and grapefruit for the European market. Turkey, Lebanon, Syria, Cyprus, and Iran, all have some commercial growing of oranges and lemons.

Along the southern coast of the Mediterranean, Egypt is second only to Mexico in lime production. Algeria and Morocco both produce large quantities of oranges for export. At the lower end of Africa, South Africa has developed an important export industry in oranges and to a lesser extent in grapefruit and lemons. Rhodesia has a small orange industry, too.

On the far side of the world, Australia has commercial production of oranges, lemons, and grapefruit, and so does New Zealand on a smaller scale, although both countries import some citrus to fill their domestic needs. All the islands of the Pacific grow various citrus fruits in abundance, but not on a commercial scale except in a few instances.

Table 3 shows the known production of different kinds of citrus fruits in the countries of the world for which records are available.

2. The Kinds of Citrus Fruits

The Orange Subfamily

CITRUS fruits and their relatives, close and distant, constitute the Orange subfamily, Aurantioideae, of the great and diverse Rue family, the Rutaceae. This subfamily is entirely Old World and is mostly limited to the tropics and subtropics of southeastern Asia and the southern Pacific, although a few genera are African. The fruit in this subfamily is a specialized berry called a *hesperidium,* usually with a leathery rind but with a hard woody shell in one small group of genera. Leaves and the bark of twigs and young branches contain oil glands, as do the leathery rinds of some fruits.

This subfamily, Aurantioideae, was divided by the late W. T. Swingle, eminent citrus taxonomist, into two tribes—Clauseneae and Citreae, the latter including Citrus and Citrus-like genera. The tribe Citreae is further divided into three subtribes of which the Citrinae contains the Citrus Fruit Trees. This in turn is separated into three groups: Group A, the Primitive Citrus Fruit Trees (including the ornamental shrub *Severinia*); Group B, the Near-Citrus Fruit Trees; and Group C, the True Citrus Fruit Trees. This last group contains six genera which possess the typical citrus fruit characters of leathery rinds with oil glands and of pulp vesicles filling the interior of the fruit. Three of the six genera are of little or no commercial importance while the other three include all citrus fruits grown commercially.

The three unimportant genera occur only in Australia and Melanesia. *Eremocitrus,* native to southern Queensland and northern New

South Wales, is the only distinctly xerophytic form in the whole subfamily and is called the desert-lime because it thrives in semiarid regions. It has great tolerance for saline soils and may be of value as a stock for citrus trees in areas plagued by high salt content of the soil. *Microcitrus* is also native to eastern Australia but has one species in adjacent New Guinea. Some species show some adaptation to low rainfall but others are found in areas of high rainfall. The fruits are known as wild-limes, being quite acid. Some of the species may have value as stocks for commercial citrus fruits. The third genus, *Clymenia*, is known only from the island of New Ireland, in the Bismarck Archipelago northeast of New Guinea, and has been little studied hitherto.

The three genera of commercial significance are *Poncirus*, the trifoliate-orange; *Fortunella*, the kumquat; and *Citrus* itself. These will be discussed in more detail below. All these genera possess certain characters in common:

1. The plants are thorny shrubs or trees with fragrant white flowers.

2. The leaves are compound in nature, still retaining this visibly in *Poncirus* with three leaflets, but reduced to the single terminal leaflet in the other genera. While appearing to be simple leaves at first glance, their compound origin is shown by the joint where the blade attaches to the petiole. Except in *Poncirus*, also, the leaves are persistent for two years or more, making the trees evergreen.

3. The petiole is often bordered lengthwise by bladelike extensions called wings. The presence or absence of these wings, and their size and shape when present, are characters useful in identifying species.

4. The typical citrus fruit, the hesperidium, is spheroidal or oblong in shape, with leathery rind which is green when immature and at maturity is green, yellow, orange, or red. This rind possesses abundant oil glands, developed by the separation and breaking down of some of the subepidermal cells. The oil contained in these glands at fruit maturity is an essential oil and differs characteristically between species. Oil from the rind of orange, lemon, and lime is commercially important. The inner portion of the rind is a whitish, spongy material known as *albedo*, while the outer, colored portion, containing oil glands and color bodies, is termed the *flavedo*.

5. The interior of the fruit, within the rind, is divided into several segments by thin membranous walls, the number of these segments having some diagnostic value at times. Each segment is packed full of juice vesicles except for the space occupied by the seeds (when present), and these vesicles have thin, easily ruptured walls normally.

6. The juice in the vesicles contains sugars, organic acids (including ascorbic acid, or Vitamin C), coloring matter, glucosides (which are

often characteristic of species), carotene (forming vitamin A) in some species, and inorganic salts, as well as traces of other organic substances. Taken together, these constituent compounds are called the soluble solids of the juice; sugars form by far the largest part of the soluble solids.

7. Seeds often contain embryos formed from the nucellus in addition to the usual embryo formed by fertilization following pollination. These nucellar embryos are derived wholly from the mother plant and reproduce it as certainly as buds or grafts do. Apparently the stimulus of fertilization is needed for them to begin development, but the sexually derived embryo may fail to develop in many cases.

8. Both bud and seed mutations are known to occur with some frequency, giving rise to "sports" which possess new and distinctive characters. Sometimes these are desirable changes and give rise to valuable new forms.

9. The species of these genera hybridize freely with one another. Many hybrids have doubtless been formed naturally during thousands of years of wild development, and they often make grave problems in estimating relationships among naturally occurring forms. Many other hybrids have been created deliberately by man, some of them having commercial value.

10. Budding and grafting are easily done reciprocally among these genera, so that a wide range of stock-scion combinations is possible. Many have been used for a long time and others seem certain to be tried in the future.

The Genus Poncirus

The genus *Poncirus* was separated in 1815 by Rafinesque from the genus *Citrus*, but the Linnaean name, *Citrus trifoliata,* continued in use by all citrus writers until Swingle called attention in 1915 to the need of transfer to another genus. It has only one species, *P. trifoliata,* and is distinguished from all other true citrus fruits by being deciduous and having 3-foliolate leaves. It is also distinct in having flower buds formed during the summer previous to their spring opening, as is common in apples and peaches, and in having the fruit covered with fine hairs. Since all other citrus fruits and their close relatives are tropical or subtropical, it seems likely that this species is derived from a primitive tropical form which migrated northward untold millenia ago, when the climate was warmer, and managed to adapt itself to the change of climate by developing the deciduous habit. The 3-foliolate leaf is undoubtedly a very primitive character, and it is somewhat surprising that no other species have been differentiated during the extended period it has been in existence in northeastern China. Long cultivated as an ornamental tree in China and in Japan, which received it in the eighth century, it came to the United States in 1869 as an introduction by William Saunders for the U.S. Department of Agriculture. Such is its hardiness that it can be grown in the open as far north as Washington, D.C., enduring temperatures as low as 0°F. when dormant. This cold-hardiness seems to be a dominant character, transmitted to hybrids of it with more tender species. It is unfortunate that the name "orange" was attached to this species on its introduction to this country, as the fruit bears only the most superficial resemblance to an orange. The pulp is not only acid but also contains a very bitter oil which makes it quite inedible, although both immature and mature fruit were utilized (along with powdered rhinoceros horn) in classical Chinese medicine. The fruit itself is small, round, yellow at maturity, with a distinct pubescence and numerous seeds.

The trifoliate-orange is known in two very similar forms, distinguished only by the size of the flowers as large-flowered and small-flowered. The species finds commercial use as a stock, either directly or in its hybrid progeny, and has minor significance as an ornamental and as protection for wild life.

Poncirus trifoliata has been valued in Florida since 1892 as a stock for citrus fruits which is not only able to endure uninjured as much cold as ever is known in this state but is also able to confer some degree of cold-hardiness on the scion top. In Florida, as in Japan, it is the only stock that is regularly used for satsumas, which are greatly dwarfed by it, and the one chiefly used for kumquats, oranges, mandarins, or grapefruit in the northern part of the state. It is now being used in a limited way as a stock in the central citrus area of the state. Scion varieties on this stock apparently are hardier to cold because of the tendency of this stock to keep them dormant during the period it would be deciduous. When tops are not dormant they are no hardier than on other stocks, which is a distinct disadvantage in years of early cold weather following a warm fall period. The trifoliate-orange is subject to exocortis (scaly butt), a virus disease which seriously impairs the vigor of the stock.

This species readily forms hybrids with species of *Citrus*, and difficultly with those of *Fortunella*. Most important of the hybrids are those resulting from crossing trifoliate-orange with sweet orange. These were first produced with certainty as hybrids by Swingle in 1897, working at Eustis, Florida, and named "citranges" by Webber and Swingle in 1905. The citranges usually come true from seed because they rarely develop a true embryo but form several nucellar embryos. They show quite a range of variation in degree of deciduousness, of leaf compounding, and of cold-tolerance, being somewhat intermediate between the two parents in these characters. Usually they are tardily deciduous, trifoliolate, and hardier to cold than sweet orange.

Citranges inherit the acidity of trifoliate-orange but have less bitterness, so that with liberal addition of sugar they supply a very good substitute for lemonade in regions further north than lemons can grow. The fruits are juicy and presumably as rich in vitamin C as lemons. The varieties 'Rusk', from the first crosses in 1897, and 'Troyer', from a cross by Swingle in 1909 at Riverside, California, have attracted attention as possibly valuable stocks and are being extensively tested. The 'Carrizo' citrange is now (1973) the stock grown in largest quantities in nurseries to replace older types which have fallen into disfavor due to disease complexes. Some other citranges are also being investigated as stocks.

The trifoliate-orange has been used as an ornamental for centuries in northern China and Japan. The white flowers and yellow fruits are rather showy, and the twigs and thorns are dark green when the leaves have fallen, making the tree attractive at all seasons. The heavy armament of thorns makes a hedge of trifoliate-orange quite impenetrable, thus affording good protection, but these same thorns may be a serious hazard if the plants are used around a playground area. On the other hand, the thorns help make this an excellent species for wildlife protection. Thickets of trifoliate-orange trees give splendid cover for ground-nesting birds, rabbits, and other animals, and have often been planted as game refuges by wildlife associations.

The Genus Fortunella

Cultivated in China for at least a thousand years and in Japan for several centuries, the kumquats somehow escaped notice by early European plantsmen to a large extent. Ferrari, in his account of citrus fruits in 1646, mentioned them as Chinese fruits on secondhand acquaintance, and Kaempfer included them briefly in his 1712 account of the plants cultivated in Japan, but no introduction to Europe was made until 1846. In that year Robert Fortune brought a plant of the oval kumquat back to England from China, where he had been collecting plants for the London Horticultural Society. It may be that attempts had been made earlier to introduce the kumquat by seed, but it does not thrive on its own roots usually and such introductions may have died before they fruited. Thunberg named the common oval-fruited species *Citrus japonica* in 1784, and this name continued in use until 1915 when Swingle decided that it was too different from all known species of *Citrus* to fit in that genus and erected the new genus *Fortunella* in honor of its introducer to the Western World.

A specimen of this kumquat was brought to America within three or four years of its reaching England, and probably was introduced to Florida very soon afterward. In 1885, however, both the Glen St. Mary Nursery and the Royal Palm Nursery made importations directly from Japan of the oval and small round kumquats under their Japanese names, 'Nagami' and 'Marumi', respectively, and most of the specimens growing in Florida derive from these introductions.

Kumquats are shrubby evergreen trees, rarely over 10 feet high, with dense branching and small, apparently simple leaves with hardly any wings. Most kumquats are nearly or quite thornless. On trifoliate-orange stock they are dwarfed; the above height is for trees on the vigorous 'Rough' lemon. The fruits are small, ovoid or globose in shape, less than 2 inches long or wide, with very thick, sweet rind and mildly acid pulp. The seeds are distinctive in having green cotyledons, whereas they are white in most other citrus fruits. Kumquats are hardier to cold than any species of *Citrus*, being able to endure temperatures down to 10°F. when fully dormant, especially on trifoliate-orange stock. In part this cold-hardiness is due to the fact that they have less tendency than do *Citrus* species to start growth during warm periods of the winter. Another factor is the delay in blooming until late in the spring, long after danger of frost is past. They are native to southern China, probably, but are known only in cultivation.

The fruits are very showy, being borne in large numbers and brilliant orange in color. The name "kumquat" is an English form of the Chinese words for "golden orange." No citrus fruit, except possibly the calamondin, exceeds it in ornamental value, and the kumquat is slower growing and more suitable to landscape usage as a specimen shrub. Since the fruit color is fully developed by December, the kumquat is well suited for use as a living Christmas tree.

Uses of the fruit are chiefly for decorating gift boxes of oranges and grapefruit and for making preserves, although the fresh fruits are quite palatable raw. For gift-box usage, twigs are cut with both leaves and fruit, the oval kumquat being preferred because it is more showy. Judicious pruning methods are used by gift-fruit shippers to assure continued production of fruit. Kumquats are excellent for making marmalade or for making candied fruits, but most varieties are quite seedy and the seeds are somewhat of a drawback for this usage. Oval and large round varieties are preferred for preserves, especially of the whole fruit. Kumquats are popular also as table decorations for parties and dinners.

The hardiness to cold of kumquats suggested their use in trying to develop hardy acid fruits, since lemon and lime are more tender than orange. Swingle first made such crosses in 1909, using the small round kumquat and the Key lime, and coined the name "limequats" for the resulting hybrids. Two varieties, 'Eustis' and 'Lakeland', have been sparingly planted to extend lime culture for home use into north-central Florida. They are less hardy than the calamondin, however.

Three varieties of kumquat are grown by Florida nurserymen quite commonly: 'Nagami', 'Marumi', and 'Meiwa'. 'Nagami', the oval kumquat, is considered a separate species, *F. margarita,* from the 'Marumi' or small round kumquat, *F. japonica,* though the differences are not easily noted. The large round kumquat, 'Meiwa', was not introduced to Florida from Japan until 1911. Swingle named it *F. crassifolia* in 1915, but in 1943 decided that it was a hybrid between the other two species and withdrew the specific name although it still persists in horticultural writings. 'Nagami' is the most popular kumquat, being vigorous and prolific. Its fruits are oval, from 1¼ to 1¾ inches long and about two-thirds as wide, with 2 to 5 seeds, and pleasantly flavored. 'Meiwa' has globose fruits, from 1 to 1½ inches in diameter, often nearly seedless, with much thicker rind than the other two forms and a somewhat sweeter taste. 'Marumi' fruits are also globular but smaller, rarely over 1 inch across, with 1 to 3 seeds. The flavor is good, but often considered a little less pleasing than 'Nagami', the rind is thinner, and the trees slightly thorny instead of practically thornless. However, 'Marumi' is a little hardier to cold than 'Nagami', which is slightly hardier than 'Meiwa'. 'Nagami' is the form best known in China, while 'Meiwa' and 'Marumi' are known only from Japan, although undoubtedly introduced there long ago from China.

The 'Hongkong' wild kumquat, *F. hindsi,* is not cultivated in Florida, although specimen plants exist in the state. It is worth growing for its bright scarlet-orange fruits, which are hardly over ½ inch in diameter. The shrubby tree is spiny and the abundant fruits are palatable only when cooked with sugar, but have a spicy flavor.

The Genus Citrus

Commercial production of citrus fruits is limited to certain species of the genus *Citrus,* which had its primary center of origin in northeastern India and a secondary center in southern China, but became widely diffused throughout southeastern Asia and the adjacent island chains. An amazing diversity of forms has developed in this area over a period of thousands of years, and authorities differ dramatically regarding which of these forms are to be considered species. Just as in oaks, blackberries, and hawthorns in the United States, the bewildering number of different types undoubtedly has arisen as the result of much crossing between original species. Most botanists recognize these resultant forms as species, for they have been able to perpetuate themselves naturally. Most of the varied forms of *Citrus* are known only as cultivated plants. Swingle* very conservatively accords species status to only 16 of these forms, and considers all the others to be hybrids of varying complexity. Yet very few of his accepted species are known as wild plants and two or three of them are almost certainly of cultivated origin, so that he is not consistent. Tanaka** very liberally recognizes 145 species of *Citrus*, accepting in this category most (but by no means all) distinctive forms regardless of the theoretical possibility of a hybrid origin, so long as this is not positively proven. Undoubtedly Swingle is much too narrow and Tanaka much too broad in assessing the number of forms that deserve specific rank. There is need of much more taxonomic study in this very complex and confused genus. The authors of this work accept some of Tanaka's species as valid, although by no means all of them.

Characters

The following characters are common to all *Citrus* species:

1. Trees evergreen, small to medium in size, more or less thorny.

2. Leaves 1-foliolate, the petioles with or without wings, the blade jointed to the petiole (except in the citron).

3. Flower buds without protecting scales, formed just prior to the advent of a growth flush, normally in early spring, sometimes in summer.

*Walter Tennyson Swingle spent 60 years studying the citrus fruits for the U.S. Department of Agriculture, and was recognized as the leading world authority. He had no peer in his knowledge of citrus relatives and in his production of hybrids by interspecific and intergeneric crosses in *Citrus* and related genera. The rich variety of forms so produced, however, led him to suspect nearly all the forms of citrus fruits found wild or cultivated of being hybrids—on purely theoretical and often ingeniously suspected grounds—and to deny species status to them on this account. Particularly in the mandarins was he blinded by this prejudice from noting valid differences.

**Tyozaburo Tanaka has also spent a long life studying citrus forms, and has investigated them intensively all across southern Asia. His knowledge of these forms from firsthand study is unrivalled. For many years a protegé and colleague of Swingle, he has opposed him bitterly since 1930 for his ultra-conservatism in citrus taxonomy. Tanaka makes a particularly strong case for recognizing several species of mandarins, instead of only one.

4. Flowers usually large, fragrant, white mostly but in a few species white tinged with pink or purple; both stamens and pistil present in the same flowers normally.

5. Fruits small to very large, with the typical leathery rind, the color at maturity varying from yellow through orange to a deep orange-red (in tropical areas the fruit may be fully mature but still green); the pulp varying from insipidly sweet to very acid, with droplets of oil in the juice vesicles, which are acrid in some species but not in any of commercial interest; segments 8 to 18, usually 10 to 14.

6. Seeds none, few, or many (there may be from 4 to 12 in each segment), the cotyledons usually white, but pale green in the mandarin group.

7. Resistance to cold variable, with definite specific differences in hardiness when fully dormant. In order of decreasing hardiness to cold the sequence is: sour orange, mandarins, sweet orange, grapefruit, lemon, lime, citron. The big problem is the strong tendency in a climate of highly fluctuating winter temperatures to resume growth prematurely. All species are very easily hurt by cold when in active growth.

Uses

The various *Citrus* species are suited to a variety of uses around the home, although by far the vast majority of plantings are made for commercial utilization. All species are handsome in foliage, flowers, and fruit, and may be used in landscaping where any other evergreen tree of similar size would be suitable. The useful fruits are an added bonus, and may be consumed fresh or made into various types of preserves, marmalades, crystallized fruits, etc.

Commercial citrus production until 1920 was concerned only with fruit for utilization fresh. Canning of citrus fruit began in 1921 and ultimately utilized a considerable part of the grapefruit crop as well as a small part of the orange crop. The advent of frozen concentrate processing in 1945 led to spectacular changes in the disposition of harvested oranges, limes, lemons, and to a less extent, mandarins. In addition to pleasant taste, citrus fruits have valuable dietary properties which have long since taken them out of the luxury class.

The processing of citrus fruits created a problem of what to do with the residue of rind, expressed pulp, and seeds. This problem has been brilliantly solved by research workers, so that valuable by-products are now derived from these former wastes. The principal by-products are dried citrus pulp (actually consisting mostly of rind) for cattle feed, molasses from the waste juice, and citrus peel oil from the oil glands of the rind; less important are citrus seed oil, pectin from the inner rind (albedo), bland syrup, and feed yeast.

Besides the long cultivated species which constitute most of the industry, a number of man-made hybrids are being cultivated now, some on a commercial scale. These are especially attractive to the small grower with a mail-order business or to the citrus hobbyist who may not engage in commercial production.

The sweet orange, grapefruit, mandarins, lemon, and lime are cultivated commercially in Florida in that order of importance. Citron, shaddock, and sour orange are grown commercially in other parts of the world for their fruit, but not in Florida. The characteristics of the various commercially important species of *Citrus* and their varieties will be considered in turn, and also the varieties of hybrid origin which are grown in Florida.

The Sweet Oranges

Both in Florida and in the world generally, the sweet orange, *C. sinensis*, is the leading citrus fruit, far exceeding in value all other citrus fruits combined. Indeed, the orange comes second only to the apple as the leading fruit in the world. The typical tree is upright in growth with branches nearly horizontal, of medium large size except on dwarfing stocks. The leaves are medium in size among *Citrus* species, the petioles with rather narrow wings which may flare out somewhat at the upper end but never overlap the blade. Fruits vary from oval to flattened-globose, with thin, smooth rind, tightly appressed, not bitter in taste, and with the pulp solid without a hollow core. Rind color ranges from light to deep orange at full maturity with cool temperatures, but with warm weather the rind may remain green or become green again if previously orange.

The great popularity of the sweet orange may be attributed to:

1. The pleasing flavor resulting from satisfactory balance of sugars and acids. Fully mature oranges are of dessert quality in competition with any other fruits, yet do not cloy the appetite of the consumer.

2. The relative ease of production of the fruit, within the climatic and soil limitations, and the possibility of both satisfactory yields and a choice of varieties covering a very long marketing period.

3. The suitability of the fruit, not only for fresh use, but for a variety of types of processing, including frozen concentrate juice, canned single-strength juice, frozen and canned sections, frozen salads, marmalades, wines, etc.

The sweet orange will remain the principal citrus fruit in Florida in the future as it has always been in the past. Almost three-fourths of the new plantings of citrus trees in this state are of this species, in large part because net returns to the grower average higher for sweet oranges than for other citrus fruits. The name "sweet orange" is used advisedly, because while "orange" means the fruit of *Citrus sinensis* to most people, it is often applied to fruit of other species with more or less resemblance to sweet oranges. The unfortunate use of the term "trifoliate-orange," in which the resemblance is definitely minimal, has already been mentioned. More confusing are such terms as "Temple orange," "King orange," "Orlando orange," "satsuma orange," "mandarin orange," and others, all of which are usually accepted as "oranges" by the general public and compete with sweet orange for the consumer's dollar, but none of which belongs to the species *Citrus sinensis.* Only the term "sweet orange" refers unequivocally to this species and its varieties.

Three groups of sweet oranges are recognized when the varieties are classified by fruit characters: normal, navel, and blood oranges. The normal, or common, orange has the characteristics cited for the species, without navel development or red pigmentation in the flesh, and its varieties constitute by far the largest part of the Florida citrus industry. Navel oranges are distinguished by fruit showing a peculiar development at the stylar end (actually an aborted second ovary), termed a "navel" for obvious reasons. Blood oranges develop red coloring throughout the flesh, or in streaks, under certain climatic conditions, and often show red color of the rind too. Under Florida conditions the blood oranges may not develop much, if any, red color in mild winters.

The Florida citrus industry also divides varieties into three groups based on season of maturity: early, midseason, and late. Early oranges are those maturing from late September through November; there are about 11 varieties in this group. Midseason oranges mature in December and January; there are around 30 varieties. Late oranges mature from February through August; only about 7 varieties are in this group. The number of varieties grown in the state has been much higher than is needed or desirable for usual commercial marketing purposes. The above numbers include all varieties grown on even a small commercial scale; only a few varieties are of large commercial importance. A great many more varieties have been selected and named in the past, most of which have been discarded entirely. For example, only two or three of some 20 varieties listed in 1880 are still being grown on even a small scale. There are also many variant types, producing fruit slightly different from the variety proper—or sometimes markedly "off-type"—which are marketed under the variety name, although they constitute only a small part of the total crop. Citrus trees are quite subject to bud mutations, and these may be propagated by the unwary nurseryman, who checks only to be sure of the variety and not to see if the fruit is characteristic.

Sweet oranges are also classed as seedy and seedless, the latter term including varieties with only a few seeds (commercially seedless) as well as those with no seeds (strictly seedless). If there are more than 10 seeds in any fruits of a variety it is considered seedy; usually the fruits classed commercially as seedless have 6 or fewer seeds.

The important commercial varieties of Florida, in order of current popularity as indicated by nursery sales, follow:

Importance	Variety	Fruit Group	Season of Maturity	Seed Condition
Major	'Valencia'	normal	late	seedless
	'Pineapple'	normal	midseason	seedy
	'Hamlin'	normal	early	seedless
Secondary	'Parson Brown'	normal	early	seedy
	'Lue Gim Gong'	normal	late	seedless
	'Summerfield'	navel	early	seedless
Minor	'Summer' (Pope, Glen)	normal	late	seedless
	'Delicious'	normal	early	seedy
	'Dream'	navel	early	seedless
	'Washington' (Glen)	navel	early	seedless
	'Ruby'	blood	midseason	seedy

These varieties are described in some detail a little later.

Any variety which supersedes one of the above must represent a distinct improvement upon it, since the large investment in the established varieties creates a hurdle which a new variety cannot leap if it is only slightly superior in some respects. Florida still has too many commercial varieties and whoever proposes to add to the number must offer a great deal more than mere novelty. Yet there is a place, for example, for a good navel variety bearing heavy crops regularly, or for a midseason variety of 'Pineapple' quality and productiveness that is commercially seedless.

In selecting varieties for large-scale planting, the grower is usually guided by the market returns for these varieties in the preceding year—or still better, over a period of years. Varieties which year after year bring good prices are those which have found favor with the public and will probably continue to do so. Another factor of equal importance is, of course, the size and regularity of the crop of these varieties, for total returns depend both on price per box and number of boxes of fruit per acre. There is one uncertainty that can hardly be guarded against, but which is always present. The variety or varieties which are profitable this year may not be the most profitable in ten years, when this year's planting comes into full production. The needs of the processing industry are not the same as those of the fresh fruit industry, although this difference has hitherto had its effect on cultural practices rather than on varieties. Even with this uncertainty, it is safer to plant varieties with proven consumer acceptance than to plant on the basis of personal satisfaction with the performance of a variety.

In selecting new orange varieties, the demands of both fresh-fruit shippers and processors must be kept in mind. The commercial packinghouse and the gift-box shipper desire oranges with attractive color of both rind and flesh, smooth and thin rind, rather fine flesh texture, pleasantly tart flavor, good shipping behavior, and fruit sizes from 150 to 250 in the standard citrus box of 1⅗ bushels. The canning and concentrating industries care little about rind color, texture, or fruit size, but want high content of juice with good color, high content of solids, fair degree of acidity, good aroma, and freedom from any off-flavor.

Orange Varieties

'Valencia'—This is the leading variety in both Florida and California, and the only orange of major importance in both states. It is not only the most important orange variety in the world, but the world's most valuable single fruit variety also. While all oranges have an ultimate origin in China, the immediate origin of this variety is uncertain beyond the fact that it was sent out by the Rivers Nursery in England around 1870. Since Rivers obtained his varieties from Mediterranean sources and this variety has long been cultivated in Spain and Italy, it is quite probable that he obtained it from one of these countries, but we cannot be sure. Attention was first drawn to it in this country in 1877 when E. H. Hart of Federal Point, Florida, announced that he had a tree with unusually late maturing fruit. The tree had been imported from Rivers Nursery by S. B. Parsons of Parsons Nursery, Long Island. He had an orange grove and nursery at Federal Point and planted it and several other varieties there in 1870. When this tree came into bearing, Mr. Hart noticed its unusually late season and found it had no label of identification. The Pomological Committee of

the Florida Fruit Growers Association examined specimens of the fruit, confirmed its remarkable lateness of maturity, and named it 'Hart's Tardif' ("tardif" meaning "late").

Another importation, also unlabelled, was made from Rivers Nursery to California in 1876, and when this tree had attracted attention, it was recognized by a visiting Spanish citrus expert as the variety known in Spain as "naranja tarde de Valencia." The name 'Valencia Late' was soon adopted (1887) in California, but it was not until 1914 that 'Hart's Late' and 'Valencia' ceased to be considered as different varieties in Florida, although many citrus authorities in California considered them identical by 1893.

'Valencia' has prime quality in Florida from late March to June, and remains on the tree in fairly good condition all summer although considerable dropping of the fruit takes place, along with regreening of the rind and drying of the flesh. There are usually 5 or 6 seeds. The quality is excellent. A number of strains of 'Valencia' have arisen by bud mutation and not all are equally desirable. No strains are propagated by name in Florida, but several have been recognized and named in California.

'Pineapple'—This is the leading midseason variety and is of even more obscure origin than 'Valencia'. About 1860 the Rev. Dr. J. B. Owens moved from South Carolina to Florida and settled near Sparr, a few miles below Citra. He planted orange seeds, which W. J. Crosby said he had obtained in Charleston from an English vessel, probably from China. When Rev. P. P. Bishop was looking for desirable trees from which to bud wild sour orange trees around Orange Lake (near Citra) in 1873, he was told of the Owens trees and bought the tops of nine trees which Dr. Owens wanted to move. One of these trees had been named "the pineapple tree" by one of the Owens girls, and when Mr. Bishop's budded trees came into fruit, the budlings from this tree had fruit superior to the others. He budded the rest of his large tract of sour oranges from this 'Pineapple' variety. There is disagreement among men who knew the original tree at the time it was first propagated as to the origin of the name "pineapple" in the Owens family. One says it was because the fruit had pineapple flavor; another that the aroma of the fruit, not its flavor, was like pineapple; and a third says it was the shape of the tree, with cylindrical sides and bushy top, which resembled a pineapple fruit. Since no one seems able to detect any flavor except that of orange in the fruit, and the marked aroma is noted only in a packinghouse full of this variety (such as the Owens never had), the third explanation is the most reasonable. The 'Pine-

apple' oranges from the Bishop Hoyt (later the Crosby-Wartmann) grove found good market reception, but only slowly were propagated by the other growers. After the freeze of 1894–95, however, when extensive replanting had to be done, this variety was widely budded. It has prime quality during January and February and has a very fine and distinctive flavor when well grown and properly mature. The number of seeds varies from 10 to 20.

'Hamlin'—This is the principal early orange of Florida and is of Florida origin. The parent tree was a seedling in a grove planted in 1879 by Judge Isaac Stone a few miles northwest of DeLand. Later this grove was bought by A. G. Hamlin while he lived in DeLand, and so the variety was propagated under his name when extensive budding was undertaken following the "big freeze" to take advantage of its unusual earliness of maturity. Its season is from late October to December, and the seeds are up to 5 in number, often none. While the quality never equals 'Pineapple', it has a notably smooth, thin rind and as good quality as any early orange. Under comparable conditions, the 'Hamlin' variety produces higher yields per acre than any other variety of sweet orange, although the individual fruits are often too small for good fresh market acceptance. Consequently it is planted about as extensively as 'Pineapple' is.

'Parson Brown'—Before 1920 this was the leading early orange, but it has given place to 'Hamlin' largely because of the former's seediness and some uncertainty of quality. The original tree of this variety was one of five seedlings at the home of the Rev. Nathan L. Brown near Webster, Florida. These had been given to him as small seedlings in 1856 by a man who said they grew from the seeds in an orange brought to Savannah from China on an English ship. When Capt. J. L. Carney in 1874 was looking, like P. P. Bishop, for a source of budwood for the wild sour orange trees on his island in Lake Weir, the fruit of one of the Brown trees took his special favor because of its fine quality before fruit of the other four trees was mature. He bought the rights to the budwood of this tree and propagated it as the 'Parson Brown' variety. The other four trees were the source of budwood for many groves following the freeze of 1894–95, because it was believed that all of Parson Brown's trees were the same, and this gave rise to much variation in the fruit and trees sold as this variety. The Carney strain represents the true 'Parson Brown' variety and matures first in October and November. It has from 10 to 20 seeds and a thick, slightly pebbled rind which loses the green color very slowly. The quality is good.

'Lue Gim Gong'—At the time of its introduction in 1912, this was believed to have originated as a hybrid of 'Valencia' and 'Mediterranean Sweet', made at DeLand in 1886 by a Chinese gardener after whom it was named. It was also considered to be even later maturing than 'Valencia'. Few would claim today to be able to tell any difference between 'Lue Gim' and 'Valencia' in season of maturity or in fruit or tree characters, and it is now considered simply a nucellar seedling of 'Valencia'. The variety is properly the 'Lue Gim Gong' strain of 'Valencia', if at all different, and not a distinct variety, though still propagated a little as a variety.

Navels—Several varieties of navel oranges are propagated and planted in Florida, but planting of unnamed navel types considerably exceeds that of named varieties. Most of them—named or unnamed—are of the same general type as the famous 'Washington' variety grown so extensively in California, but none of them is anywhere nearly so important—nor are all of them taken together—in Florida as 'Washington' is in California. Navels as a group make up less than 3 per cent of the orange trees in Florida, yet the fruit often commands a premium early in the season as a "salad" fruit. 'Washington' itself was introduced to Florida soon after it proved so successful in California, whither budded trees were introduced from Brazil via Washington, D.C., in 1873. Indeed, some trees seem to have been sent by the U.S. Department of Agriculture directly to Florida at the same time as to California, but real interest developed only after the California success, from 1880 on. This variety proved a very shy bearer in Florida, however, although quality was good, and so has never been extensively planted, although still propagated on a minor scale. 'Washington' tends to make very large, coarse fruit when the crop is very light, often with early dryness and an unsightly navel.

'Summerfield' is the navel variety most popular currently in Florida. Several old trees have been found in groves south and southwest of Lake George on the St. Johns River, and were probably derived from nucellar seedlings from the navel orange introduced from Brazil in 1835 by D. J. Browne and planted for him on Drayton Island by Zephaniah Kingsley. This was the same Bahia navel type introduced and named 'Washington'. Studies of 'Washington' in California have shown that a large number of slightly different strains have arisen there by bud mutation. 'Summerfield' may represent a strain of the same original stock as 'Washington', but one better adapted to Florida conditions, so that it bears somewhat heavier crops, and fruit size is

smaller in proportion to crop. The fruit is otherwise very like that of 'Washington'. Summerfield Nursery Company had the parent trees brought to its attention and has been propagating it commercially since 1929. Some fairly large plantings were made from 1955 to 1960.

'Glen Improved' represents another strain, this time derived from 'Washington' itself. The parent trees were noted in a planting of 'Washington' in a Polk County grove of Wm. G. Roe of Winter Haven. These trees, evidently budded from an unrecognized mutant branch, were bearing good crops of normal-sized fruit. Since 1934 this strain has been propagated commercially by Glen St. Mary Nurseries Company as 'Glen Improved'. Good results are obtained only on heavier soil types where sour or sweet seedling stocks can be used successfully.

'Dream' is a navel variety which was found as a budded tree in a grove in Seminole County by D. J. Nicholson in 1939. In 1944 he took out plant patent No. 625 on this variety and began commercial propagation. It undoubtedly represents another bud mutation from 'Washington', although an unusually favorable one. According to the originator it matures in early October, which is quite early for a navel, and can be held on the tree for several months without drying out—whereas the large scanty fruit of the typical 'Washington' tends to dry out badly in a couple of months after being mature. Like the 'Glen Improved', the variety is not very successful in deep sandy soils on 'Rough' lemon stock, but is a beautiful and high-quality fruit when properly grown.

'Pope Summer' ('Glen Summer')—For over 60 years varieties of orange which can be held on the tree without excessive dropping a little later than 'Valencia' and 'Lue Gim Gong' have been marketed as Summer oranges. They are always similar to 'Valencia' and may represent mutant strains. The parent tree of the 'Pope Summer' was found in a 40-acre planting of 'Pineapple' near Lakeland by F. W. Pope about 1916. Evidently one tree was budded in error. The contrast between the very late season of maturity of this one tree and the surrounding 'Pineapple' trees was very striking, for the fruit held well into August. Commercial propagation was begun in 1935, and since 1945 has been carried on by the Pope Summer Orange Nursery, Ltd. The name 'Pope's Summer' was trademarked in 1938. On sour orange or sweet seedling stocks, especially on hammock soils, the fruit is good quality in July and August, though mature in May. The same variety is propagated by the Glen St. Mary Nurseries Company as 'Glen Summer'.

'Queen'—The parent tree of this variety was a seedling tree of unknown origin in the King grove on Lake Hancock, near Bartow, which survived the freeze of 1894–95. In 1900, buds from this tree were obtained for use in the planting of the Perrin and Thompson grove at Winter Haven, because it seemed to have fruit of unusually good qualities. At first C. H. Thompson called the variety 'King', to honor the man in whose grove the parent tree grew, but later he changed the

FIG. 2. Citrus groves in lake country of central Florida. *Courtesy Citrus Tower*

name to 'Queen' to avoid confusion with the well-established 'King' mandarin. Commercial propagation was begun by Lake Garfield Nurseries in 1915, and the variety has had a mild popularity, increasing in recent years. The fruit is much like 'Pineapple' in season and quality, with rather fewer seeds, but tends to be higher in juice solids on 'Rough' lemon stock and to hold better on the tree during periods of deficient soil moisture. The trees grow vigorously and yield well, and are slightly more hardy to cold than 'Pineapple'.

'Jaffa'—Introduced by General Sanford in 1883 from Palestine, where it was reputed to be a very fine variety, the 'Jaffa' soon became popular in Florida because of having few seeds together with good quality. After 1895 its popularity declined in favor of 'Pineapple', because of poor bearing habit, and for many years it was little planted. Since 1945 there has been a small revival of interest. The fruit ripens in midseason, is of medium size, and has from 6 to 10 seeds. Quality is very good and the yields are also satisfactory in many groves.

'Delicious' ('Partin Delicious')—The parent tree of this variety was a seedling of unknown origin which was observed by Clay S. Partin around the turn of the century. Propagation for their own groves was begun by the Partin family soon after 1900 and has continued ever since, but commercial propagation was started by the Lilian S. Lee Nurseries in St. Cloud in 1958. The fruit has much resemblance to 'Parson Brown', but has somewhat smoother rind, deeper flesh color, and fewer seeds. Fully as early maturing as 'Parson Brown', the fruit can be held on the tree without dropping or drying out until March, at least on low, moist soils. Trees are unusually upright in habit and bear good crops of fruit with good quality and 6 to 10 seeds. Most plantings have been on sour orange stock in Osceola and southern Orange counties, but plantings in deep sandy soils on 'Rough' lemon have also produced good quality fruit.

'Enterprise'—This early, commercially seedless variety was formerly a prominent one in its season in Florida. Reputedly the parent tree, found at Glenwood (a few miles northwest of DeLand) about 1880, originated as seed from the old Dummitt grove on the north end of Merritt Island, presumably planted in 1830 with trees budded from the Mays grove at Orange Mills, which in turn was supposedly set out in 1824 with budded trees brought from Spain. The variety was named for the town of Enterprise, across Lake Monroe from Sanford. 'Hamlin' has long since replaced it as the leading early seedless variety, but 'Enterprise' continues to be planted sparingly. The tree is vigorous and bears well, and the fruit is very good quality with about 5 seeds, or often none.

'Ruby'—This is the only one of several "blood" oranges, once popular in Florida, which is still propagated commercially, and demand is small for it. General Sanford introduced it from the Mediterranean area about 1880. The quality of 'Ruby' is very good but not more so than 'Pineapple' and 'Jaffa', which mature in midseason with it. The streaks

of red color in the flesh develop only with quite cool weather, and in southern Florida many growers have been disappointed in the failure of the fruit to show the "blood" character. Fruit exposed to light on the tree has more color developed than well-shaded fruit. The variety is grown chiefly for home use and gift-box shipments, where its novel coloring combined with high quality (well-matured fruit has more color) makes a big appeal. Seeds number about 10.

Grapefruit

Grapefruit is considered a distinct species (*Citrus paradisi* Macf.), although it is known only in cultivation. It appears to have arisen, probably as a mutant of the shaddock (*C. grandis* Osbeck), in the West Indies during the eighteenth century, and was first heard of in Barbados in 1750 under the name "forbidden fruit." In 1789 the "forbidden fruit or smaller shaddock" was reported common in Jamaica, and in 1814 it was called "grapefruit" there, the name being given because the fruit hung in small clusters, like some grapes, instead of one on a twig as is common with oranges and shaddocks. The specific epithet, *paradisi*, reflects the early name "forbidden fruit." During the first quarter of the present century a determined effort was made by horticulturists to have the name "pomelo" adopted for this species instead of grapefruit, but the latter name was too well fixed in the minds of growers and the buying public. Pomelo is still the common name in Brazil, although *toronja* is the common name in Spanish, but it is practically never used anymore in the United States.

Grapefruit not only originated in the Western Hemisphere but it is little cultivated anywhere in the Old World, although Israel and South Africa have commercial plantings for export. The United States produces about 90 per cent of the world crop. The introduction of grapefruit to Florida took place in 1823 when Dr. Odette Philippe, a French count, settled near Safety Harbor on Tampa Bay, bringing with him seeds or seedlings of grapefruit and other kinds of citrus fruits from the Bahamas. So far as we know, all of our present varieties are descended from this introduction, by seedling variation, mutation, or hybridization.

The new fruit was very slow in attaining popularity, and was chiefly a curiosity until about 1885, usually being considered merely a variety of shaddock, although considerable quantities were imported from the West Indies (78,000 fruit into New York in 1874). In 1875, Florida nurseries offered only grapefruit seedlings, and a book on Florida fruits in 1886 mentions grapefruit only as the best of the shaddocks and unworthy of commercial cultivation. Yet in 1884 a New York fruit dealer was quoted in a Florida paper as saying there was a considerable demand for the fruit. Evidently some shipments were being made already, for in 1884–85 J. A. Harris reported he had shipped 1,500 barrels from his Citra grove, although in 1889 W. S. Hart considered that there was really no commercial production yet. He noted that several improved varieties were being offered by nurseries, but at the State Horticultural Society meeting that year there was a premium only for the best plate of grapefruit (no varieties) while premiums were offered for about 50 named varieties of oranges. The first named grapefruit varieties appear in horticultural literature only in the 1897 variety list of the American Pomological Society, although nursery catalogs had been listing some varieties for several years.

Florida has the largest production of grapefruit, with Texas a good second, and much smaller production in California, Arizona, and Puerto Rico.

In production and consumer acceptance grapefruit ranks definitely behind the sweet orange. World production of grapefruit is only about one-sixth the production of oranges, and Florida produces more than 50 per cent of all the grapefruit in the world. In addition Florida can claim credit for originating all the important (and many more unimportant) varieties except 'Ruby', which appeared first in Texas. The first planting of grapefruit trees in the United States, the first shipment of grapefruit to northern markets, and the first commercial plantings of grapefruit also took place in Florida.

In spite of an auspicious beginning, grapefruit growers in Florida have had many difficult years. Often, during the 1950's and 1960's, the fruit was simply left on the tree (economic abandonment) because the prevailing prices would not pay for picking, hauling, packing, shipping, and selling costs, let alone return any profit to the grower. The immediate cause of this situation was, of course, production far in excess of consumer demand; but the basic factor in small demand is that the "tonic" flavor of grapefruit is relished by a far smaller segment of the population, generally those of middle age or above, than is true of oranges. And the tendency to ship grapefruit prematurely—a tendency inveighed against at least as early as 1910—has not improved the situation. Furthermore, although canned grapefruit segments and juice proved a more satisfactory product than canned oranges, the frozen orange concentrate is far more popular than either canned or frozen grapefruit.

For many years, few plantings of grapefruit were made in the state because of the above-mentioned possibility of economic abandonment.

Since 1968, however, growers have begun to make new plantings with the realization that the older plantings of this fruit were gradually being exhausted. This took on greater proportions with the increased demand, and better prices, for grapefruit occasioned by consumer interest in the diet value of the fruit. Plantings are continuing, especially in the Indian River Citrus Area, to satisfy this recent market.

The original grapefruit trees, and the ones first planted in Florida, had yellow rinds at maturity and flesh of pale ivory-yellow, usually spoken of as white. By rather infrequent mutation this color may become pink, and the pink form may produce one with red flesh. So far as is known pink and red colors have appeared only as bud mutations, where a single bud on a tree develops into a branch bearing fruit of a color different from that on all the other branches. Buds taken from this mutant branch will produce trees with fruit exactly like this branch bore, and so the new form can be propagated as a variety. While the earliest mention known of grapefruit in Florida, by Atwood in 1867, refers to the red flesh of the fruit, all the seedlings known in the 1880's seem to have borne fruit with white flesh. It was not until 1906 that a pink-fleshed sport ('Foster') was found, and only in 1929 that one with red flesh ('Ruby') was discovered. While pink, or especially red, flesh may have eye appeal for the buyer, it should be emphasized that there is no correlation at all between color and taste. The color sports taste like the varieties of which they are mutants. From about 1935 there was quite a vogue in Florida for planting pink, seedless grapefruit, followed about 1945 with similar heavy planting of red, seedless forms. Since 1954, however, there has been a notable decrease in enthusiasm for red grapefruit, and there has even been extensive topworking of these varieties to other citrus forms, sometimes even before they had begun to bear. Since 1955 more Marsh trees have been planted than any other variety. Now, 1973, the new plantings are of the various colored types. With the advent of the 'Star Ruby' in Texas, there is considerable interest in its introduction, through the proper authorities, into Florida.

The seedless condition, by which is meant that very few or no seeds develop beyond a very early stage, has also arisen in grapefruit by mutation, but in this case by seed mutation, not by bud mutation. This simply means that the genetic change occurred during the formation of the embryo of the seed, and that all branches of the resulting tree are alike. At least four instances are known of seeds from normal seedy grapefruit producing trees with seedless fruit, and four seedless varieties—'Cecily', 'Davis', 'Marsh', and 'Star Ruby'—have been propagated in consequence. Only 'Marsh', however, has been grown to any considerable extent, although 'Star Ruby' is being extensively planted in Texas. While 'Marsh' is now the leading grapefruit variety of the world, as well as of Florida, it is the convenience of not having seeds rather than superior quality which has put the seedless varieties ahead. It is generally conceded that the seedy varieties excel the seedless ones in taste. For canning, also, the seedy varieties are much more satisfactory than seedless, and only white flesh is desired.

The color mutation is quite independent of the mutation for seedless condition, and pink and red forms are grown of both seedy and seedless types. Since the 'Duncan' variety was a seedling of the original tree introduced to Florida in 1823, we may consider that all other varieties have arisen from it. The relationship of the color mutants to their parents is shown below.

FLESH COLOR

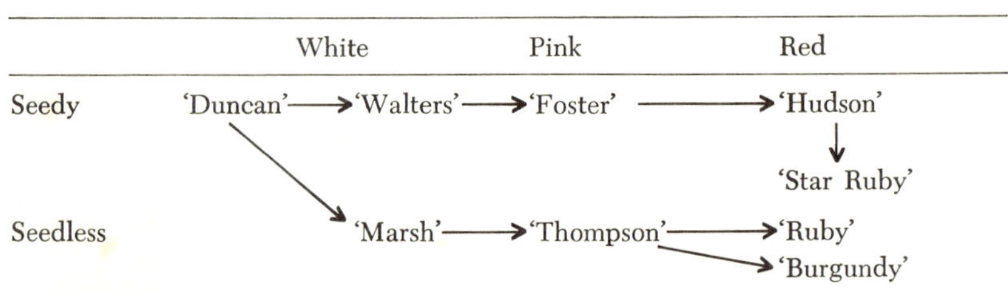

Red, seedy grapefruit are now being propagated in Florida, but no varieties have been described and the possible relation to 'Foster' (no other pink, seedy variety has been recorded) cannot be determined. The 'Hudson', a red-fleshed, seedy bud sport of the 'Foster', was named and described in Texas. From seed of this variety arose the 'Star Ruby' which is red-fleshed and seedless. In the seedless varieties the sequence is clearly established as indicated. Many of the seedy varieties with white flesh are so similar to 'Duncan' that they are practically indistinguishable, and they are all sold on northern markets as 'Duncan' or as 'Florida Common'. Some distinct varieties have arisen by natural seedling variation, since only about 9 out of 10 grapefruit seeds reproduce the parent tree exactly. Other varieties have appeared which lack the slightly bitter taste which is characteristic of grapefruit, and while seeming otherwise to be true grapefruit, are on this account considered possibly to be hybrids. These include the 'Royal' and 'Triumph' of Florida, the 'Imperial' of California, and the 'Chironja' of Puerto Rico, which appear to be a natural group of orange-grapefruit hybrids (orangelos). It is also possible that the loss of the bitter principle is due to a gene mutation.

Grapefruit varieties, like oranges, may mature early, in midseason, or late, when allowed to develop normally. Some of the seedy varieties naturally develop a pleasing ratio of sugar to acid by November, while the seedless varieties and other good seedy varieties are not naturally mature enough for good eating quality until February or March. Prior to the use of arsenic spray, the commercial practice was to consider the seedy as early and the seedless ones as late, since the latter developed the minimum solids for legal maturity later than did the seedy varieties constituting the bulk of commercial plantings. With the advent of maturity laws based on ratios of solids (sugars) to acid as well as on minimum solids, and the general use of arsenic sprays to prevent normal development of acidity, almost any variety can pass the legal test of maturity in the autumn, and 'Marsh' is shipped as early as any other variety. For the homeowner, however, who does not plan to use arsenic sprays and wants grapefruit of highest quality, it is still helpful to know the normal season of maturity of different varieties.

The different markets for grapefruit do not all want the same fruit characters. The fresh fruit and gift-box shippers demand a regularly oblate shape which can be cut into two symmetrical halves for serving at the table. Fruit should have a rind which is smooth and uniformly yellow or blushed, without any green. Seedless fruit is preferred, and pink or red flesh color is usually an advantage, although white flesh sometimes brings better prices. The market wants sweet fruit, and this can often be assured before very late in the season by preventing acidity from reaching normal values; so fruit for this market is commonly arsenated. Desirable sizes are large to medium—from 54 to 96 per standard box.

The producer of canned sections and citrus salads is concerned with having pleasant tasting fruit and segments which are intact after processing. Pink and red flesh colors do not usually look attractive after canning and seedy fruit has better section stability than seedless, as well as higher solids content; so seedy varieties with white flesh are used generally.

For canned juice and frozen concentrate the important qualities are attractive juice color, high juice content, high solids, a fair degree of acid, and low bitterness. Again the seedy, white-fleshed varieties are preferred, although much of the seedless crop is also processed.

In 1960, nearly two-thirds of the grapefruit trees in the state were seedless varieties, but only two-thirds of them were in full bearing, whereas all but about one-tenth of the fewer seedy trees were over ten years old. Consequently the seedless crop was not much larger than the seedy crop, but the difference was increasing each year in favor of seedless. By 1970, three-fourths of the grapefruit crop was composed of seedless varieties. The fresh market is made up almost exclusively of seedless varieties. As much as three-fourths of the processed products are seedless although seedy fruit are required for sectioning since the segments of seedless fruit break up upon sectionizing. The firm flesh of the 'Star Ruby' may allow it to be sectionized.

Grapefruit Varieties

'Duncan'—This is the oldest grapefruit variety, although not the first to be named and propagated, the parent tree having been planted around 1830 near Safety Harbor and having lived to be nearly a century old. About 1892 it was introduced and propagated by A. L. Duncan of nearby Dunedin. It has been the standard of quality in grapefruit ever since its introduction, and is probably the hardiest grapefruit variety to cold, but it has from 30 to 50 seeds. It matures normally from January to March, and has white flesh. As mentioned above, while 'Duncan' is still propagated and planted as a distinct variety, any seedy, white-fleshed grapefruit may sell on northern markets under this name, since it is the prototype of this common sort of grapefruit.

'Foster'—This was the first variety with pink flesh to be discovered. It arose as a bud sport of the 'Walters' variety and was first noted in 1907 by R. B. Foster of Manatee in the old Atwood grove near Ellenton. It was first propagated by the Royal Palm Nursery at Oneco in 1914. The rind is also flushed pink, so that the fruit as it hung on the parent tree was quite in contrast to the yellow fruit of the 'Walters'. The season of normal maturity is November and December. Quality is good, but the number of seeds is as large as or larger than in 'Duncan'.

'Marsh' ('Marsh Seedless')—Only one variety ('Triumph') is known to have been propagated earlier than this one. The original tree was a seedling supposedly planted about 1860 near Lakeland. It was first noticed with interest in 1886 by E. H. Tison who began propagation and offered it in 1889 for the first time simply as "budded seedless grapefruit (very choice)." A few years later C. M. Marsh bought this nursery and named this variety 'Marsh's Seedless', but the name was long ago shortened to 'Marsh'. The fruit is much smaller than fruit of 'Duncan' and 'Foster' and normally matures a little later, and it can be held longer on the tree without excessive dropping. Flesh color is white and the seeds range from 6 down to none in the best strains.

'Marsh' has long remained the leading variety of grapefruit in Florida and the world.

'Ruby'—The first grapefruit with deep red flesh originated in a planting of 'Thompson' made at McAllen, Texas, in 1926 by A. E. Henninger. When the trees first fruited in 1929 one branch had fruit with a crimson blush, and on cutting it open the flesh color was ruby red. Plant patent No. 53 was issued to Mr. Henninger in 1934, but several similar mutations of pink 'Thompson' to red flesh and rind, which cannot be distinguished from 'Ruby', have appeared in both Texas and Florida. Many nurseries now simply propagate them as 'Red Seedless'. Season and quality of these red sports is the same as for 'Thompson'. There is a tendency to call this the 'Redblush', the name widely used in Texas where the variety originated, which is more descriptive than 'Ruby' due to the blush on the rind occasioned by the red albedo. Furthermore, this name avoids the confusion with the much older 'Ruby' orange variety. This red-fleshed variety is second to 'Marsh' in current grapefruit plantings in Florida.

'Thompson'—This variety originated in 1913 as a pink-fleshed bud sport of a 'Marsh' tree in the grove of W. B. Thompson at Oneco but was not propagated until the Royal Palm Nursery of Oneco introduced it in 1924. It is seedless, just as 'Marsh' is, and of the same quality and external appearance, for no pink shows on the rind. 'Thompson' matures earlier than 'Marsh' when the fruit develops normally, but is like it in holding the fruit well without dropping for several months, although the pink color fades out with age.

'Triumph'—While not of commercial importance, this is one of the best varieties for home planting because of its unusually fine flavor and freedom from bitterness. It is very seedy but also very juicy. It was the first variety to be named and propagated, having been offered to the public in 1884. Nothing is known of its origin except that the parent tree was growing on the grounds of the Orange Grove Hotel in Tampa. The varieties 'Royal' and 'Isle of Pines' are almost or quite the same as 'Triumph', and all mature in November without arsenic treatment. As mentioned previously, this variety, as well as the 'Imperial' and the 'Chironja', may be rightly called an orangelo (orange x grapefruit hybrid).

'Burgundy'—The parent tree was found about 1948 in a grove of 'Thompson' near Ft. Pierce, probably originating as an undetected bud sport of this variety. It was patented in 1954 and introduced com-

mercially in 1956. The fruit is red-fleshed, seedless, and late-ripening with white albedo similar to the 'Thompson'. Known locally in the Indian River Citrus Area, it has not been widely planted.

'Star Ruby'—This variety was developed by the Texas A & I University Citrus Center from an irradiated seed of the 'Hudson' variety. It differs from the 'Hudson', a seedy, red-fleshed variety, in being commercially seedless with 0–9 seeds. Similar to its parent, it often shows a reddish tinge of wood and bark at the cambial layer. Leaves, narrower than those of other grapefruit varieties, often exhibit chlorotic patterns of white margins and areas although the condition is not chimeral. The tree is bushy and compact, blooms profusely, and bears some fruit in clusters. Flowers produce abundant pollen; seeds are large and polyembryonic. The original tree has produced substantial crops of good-sized fruit with red albedo. The 'Star Ruby' may find a use in the sectionizing industry since segments hold together well; it is suitable for juicing, too, since at single-strength it retains its color well. Introduced in Texas in March 1970, it was officially released in Florida on November 15, 1973. As a note of caution, some future budded trees may revert to the seediness of the parent tree.

The Mandarin Group

The name mandarin (or mandarin orange) is applied to a number of citrus fruits which have as their common characters, distinguishing them from oranges, a peel which is quite free from the flesh, green cotyledons usually in the seeds, and flowers singly or in clusters which never form a branched inflorescence. The term "kid-glove orange" was at one time applied because it was said that a lady could eat one while wearing kid gloves without getting them wet; and recently mandarins have had some publicity as "zipper-skinned oranges." They are not properly oranges at all, and the tendency to distinguish them in the public mind as mandarins instead of as oranges is increasing. The mandarin group as a whole is one of specialty fruits with rather short season and some marketing difficulties for distance shipment because of ease of injuring the soft, loose rind, with resultant development of rots and molds.

Three species are grouped under this name—the satsumas, the King and kunenbo, and the tangerines (or mandarins proper), though Swingle put them all in one elastic species, *Citrus reticulata* Blanco.

Satsumas (*Citrus unshiu* Marc.) (*C. reticulata* of Swingle)

The satsuma originated in Japan in about A.D. 1600 from seed brought from China, much as the grapefruit did in the West Indies. The Japanese name, Unshiu, is a corruption of Wenchow, the area of China from which the seeds came. It represents a mutation in all probability, again resembling grapefruit in origin, for no similar citrus fruit is known in China. In Japan the satsumas are the most important citrus fruit grown on a large commercial scale. The fruit is extensively exported to the United States and Canada, although the bulk of the crop is consumed at home. There are several varieties of satsumas cultivated in Japan, and while at least four of these are known to be in cultivation in this country, only one variety ('Owari') is at all common.

The 'Owari' satsuma was first introduced to Florida in 1876 by Dr. George Hall and in 1878 by Gen. George Van Valkenburg, both importations being of budded trees. It is notably more resistant to cold than the sweet orange and so is successfully cultivated all around the Gulf Coast from Florida to Texas. It is usually considered satisfactory only when grown on trifoliate (or citrange) stock, on which it makes a very dwarf tree, and is cultivated almost entirely north of the main citrus belt. However, there have been cases of vigorous growth and fruiting on sour orange stock. The trees bear heavy crops of fruit with distinctive flavor and attractive appearance which mature in October and November with the earliest oranges. Both rind and pulp are deep orange in color. If left on the tree more than two months after they reach maturity, the fruits become puffy and lose their sprightly taste. Normally they are seedless.

King (*Citrus nobilis* Lour.) (*C. reticulata* x *C. sinensis* of Swingle)

The 'King' orange of Indo-China and the closely related 'kunenbo' of Japan are the only representatives of this species. They may have arisen many hundreds or thousands of years ago as hybrids, but they have maintained a distinctive existence since then and have never been duplicated in any man-made crosses, so that they seem entitled to species recognition.

The 'King' was introduced as fruit from Saigon to California in 1880, and thence as small seedling trees to Florida in 1882, fruiting here first in 1884. The fruit is the largest, with the thickest rind, of all the mandarin group and is also the best in quality, commanding a premium price when well grown. It has never been important commercially because of a tendency to overbearing, with resultant breakage of limbs, and to sunscalding of fruit because of upright branches with relatively sparse foliage. Furthermore the rough rind is unattractive, with no sales appeal except to the few who know its superb flavor. However, there is a small, steady demand for nursery stock for home planting and for gift-box shipments. The season of maturity is March to June, and the fruit has several seeds which, unlike all the other seeds of the mandarin group, have white cotyledons. The fruit can be sectionized and used in salads.

Tangerines (*Citrus reticulata* Blanco)

There is no satisfactory distinction in meaning between "mandarin" and "tangerine." The former word is of older usage, and the earliest use of "tangerine" (1841) is as a synonym for mandarin. The name was originally spelled "tangierine" because fruits were imported from Tangiers in Morocco to England. At one time it was customary to separate the varieties with yellow-orange rind as mandarins and those with reddish-orange rinds as tangerines, but this does not accord with Florida practice of recent years. Yellow 'Oneco' and red 'Dancy' are alike called tangerines here. It seems desirable to let the word "mandarin" be the group name, as is widely done, for all the loose-skinned, sweet citrus fruits, and to refer to all mandarins except satsumas and 'King' as "tangerines." The authors restrict the name *Citrus reticulata* Blanco to certain tangerine-mandarins.

The first mandarins (tangerines) introduced to Florida must have come about 1825, for the village of Monroe on the St. Johns changed its name in 1830 to Mandarin and this must have been the result of pride in possessing this novel citrus fruit. We know nothing of the source or the variety thus introduced, though the 'ponkan' was available in England after 1805, nor do we know the variety of the second importation in 1838. This came from China, however, by way of Parson's Nursery on Long Island, and on its leaves came the long scale which nearly wiped out all the citrus trees of the state. Perhaps all the tangerine trees were killed, for in 1877 Major Atway was given credit for introducing this type of citrus fruit to Florida from Louisiana. The Read and Hartley groves at Mandarin survived the long scale, but they may have had no tangerine trees. But what Major Atway introduced is somewhat uncertain. Usually he is credited with bringing in the 'Willowleaf' ('China') variety, which originated under cultivation in Italy and was well known there long before 1840, the year in which Tenore described it as a new species. It is supposed to have been introduced to New Orleans from Italy between 1840 and

1850, although no basis was given for these dates by the writer who made this statement in 1902, and it could easily have been several years before 1840. The Atway grove at Palatka was bought by Dr. N. H. Moragne in 1843, and his daughter has stated that a mandarin tree was growing there at that time. This could have been the 'Willowleaf', though this is unlikely. But the fruit which the Pomological Committee of the Florida Fruit Growers Association described from the Moragne grove in 1877 as "Tangierine Orange (synonyms: Mandarin, Kid Glove, Tomato Orange)" and ascribed to Major Atway's introduction was definitely not the 'Willowleaf' but very similar to the 'Dancy'. It seems that either the Committee erred or Major Atway did not introduce the 'Willowleaf'. The Moragne tangerine was stated by Col. Dancy in 1885 to have been introduced from Tangiers about 1850, and he was a neighbor of Dr. Moragne at that time.

The only 'Willowleaf' tangerine known to have been in Florida in 1877 was one small tree in the grove of E. H. Hart at Federal Point, a few miles downriver from Palatka. This had been obtained from England by S. B. Parsons in 1870. Since this type was not known in China, it could not have been imported from China in the 1830's, and it almost certainly was the type introduced to Louisiana in the 1840's. If Major Atway brought it to Florida, it must have died before 1877, for Hart certainly knew the trees in the small Moragne grove and his tree was unique. Much mystery surrounds the supposed Atway introduction, but in any event it was definitely not the first one for mandarins in Florida. The first mention of 'Willowleaf' by name (as 'China') was by A. H. Manville in 1883, and it was Pliny Reasoner in 1887 who erroneously made it a synonym of the 'Moragne' tangerine.

The name *Citrus reticulata* Blanco was originally applied to the tangerine known in Formosa as "ponkan" and in India as "suntara." Tanaka would limit its use to this important and widely distributed form and apply separate specific names to each of the many other types of tangerines. Undoubtedly the differences between 'Dancy' and 'Ponkan' are much greater than those between 'Parson Brown' and 'Valencia', and might well justify specific distinction, for they seem comparable to the differences between 'Nagami' and 'Marumi' kumquats. Some of Tanaka's species, however, seem based on rather trivial differences, and some of our varieties do not fit in any of his species. For this reason the authors use *Citrus reticulata* to include all the tangerines, although recognizing that several good species may well be included therein.

Some tangerines are native to India, some to Malaya, several to China, and one (as above noted) is of Italian origin. All the types grown on even a small commercial scale in Florida are marketed as tangerines, but until about 1945 the only tangerine of real value was 'Dancy'. The following paragraphs are based on experience with this variety because it is the only one which has been marketed for a long period in large quantity. Some of the other varieties which have been planted commercially since 1945 give promise of being free from some of the serious defects of 'Dancy', but in 1900 nobody considered that it had any defects.

As specialty fruits, tangerines have long been popular for home use and for gift boxes. The attractive color of rind and flesh, the ease of removing the peel and separating the segments of the flesh, and the sweet taste with pleasantly piquant flavor make them desirable as dessert fruits. But as fruits for large commercial production they have several disadvantages in comparison with oranges. If left on the tree after becoming mature, the fruit suffers a rather rapid loss of acidity which leaves it insipidly sweet, and the segments dry out. With shrinkage of the segments from drying out, the space between flesh and peel increases, causing "puffiness." Tangerines tend to overbear, producing a large crop of uneconomically small fruit one year which so devitalizes the tree that only a very light crop (of desirable large fruit!) is borne the following year; and then the alternating cycle is repeated. Fruit produced inside the canopy of foliage is much less well colored than that borne on the outside, and may have poorer internal quality also. Loss of green color and development of red is often slow, and since attractive color is an important factor in selling price, it is necessary to make a series of spot-pickings for size and color. Even after several pickings much fruit may still remain on the tree, especially in a mild winter.

In the early days of the citrus concentrate industry it was hoped that tangerines of undesirably small size and poor color for fresh shipment might be utilized as frozen concentrate. This hope has not yet been fulfilled to any great extent, about 20 per cent of the crop being processed, but mostly canned rather than frozen.

The shipping season for ('Dancy') tangerines is short, the best market being in the period from Thanksgiving to Christmas for the holiday trade. There is a secondary period of demand in spring which can only be filled by late-bloom fruit. In transit, tangerines are highly susceptible to injuries which permit rapid invasion by rots and molds.

The price per box "on the tree" for tangerines is often higher than that for oranges, but the number of boxes of marketable fruit per acre is lower and the net return per acre is lower for tangerines. Many shipping organizations refuse to handle tangerines, although when sales efforts are specifically directed to this fruit, satisfactory market arrangements are possible.

At one time tangerines were very extensively planted in Florida, the only state which markets a large crop. A considerable number of older plantings have been topworked or replaced because of unsatisfactory returns, and since 1930 new plantings of all tangerines have been relatively small. Some of the citrus hybrids (tangors and tangelos) which will be discussed later have some of the desirable qualities of tangerines without some of the serious disadvantages; they are often marketed as competitors of oranges rather than of tangerines, although they should be considered specialty fruits.

TANGERINE VARIETIES

'Dancy' (*Citrus tangerina* of Tanaka)—This is the variety which is usually thought of when the word "tangerine" is heard, and until 1945 it was planted almost exclusively for commercial production. The original tree of 'Dancy' developed in the grove of Col. F. L. Dancy at Orange Mills from seed of the 'Moragne' tangerine. The variety is typical of the tangerine type native to the Foochow area of southern China and introduced to Japan in the sixteenth century probably, where it is known as 'Obeni-mikan'. This variety reproduces practically 100 per cent true from seed unless a mutation occurs, which is very rarely, and the source of the seed is known in this case. The first mention of the variety is in the report of the Pomological Committee in 1877, which considered this a new fruit similar but slightly superior to the fruit of Dr. Moragne's tangerine. Various opinions have been expressed on its origin, but Col. Dancy himself stated in 1885 that it arose from a seed of the 'Moragne' tree planted in 1867. Commercial propagation began about 1890 by the Rolleston Nursery at San Mateo. In bright color and pleasing combination of sweetness and sprightly acidulousness, 'Dancy' is typical of all tangerines, and has the typical defects also. It matures in December and January and has from 7 to 20 seeds.

'Ponkan' ('Chinese Honey' orange, 'Warnurco' tangerine) (*Citrus reticulata* of Tanaka)—This variety, as above indicated, is of Indian origin but has been popular in southern China for many centuries, and also in Formosa and Japan, where it is called 'Ponkan'. Two fruits of this variety were sent about 1892 to J. C. Barrington of McMeekin (near Hawthorne) by Dr. Parks, a medical missionary in China. He gave some of the seedlings raised from the seed of these fruits to Frank Jenkins of Melrose, and one of them survived the 1895 freeze,

as did two seedlings planted by Mr. Barrington. Dr. Parks apparently sent them as "honey oranges," and the fruit attained some small fame as 'Chinese Honey' orange. In the 1920's the Wartmann Nursery Company at Ocala propagated this fruit under the name 'Warnurco' tangerine. Under its proper name, 'Ponkan', it has been somewhat extensively planted since 1945, but has never been prominent in the market.

The fruit of the 'Ponkan' is more nearly globose than most tangerines, and is heavy and soft, with very sweet, melting pulp. The rind color is orange, not reddish like the 'Dancy', and the seeds are few. The season of maturity is November and December, and fruit left on the tree does not hold its quality long.

'Oneco'—Introduced from India as seed by P. W. Reasoner of Royal Palm Nurseries, Oneco, in 1888, this variety was identified by Tanaka in 1929 as merely a strain of the 'Ponkan'. Certainly it resembles the 'Ponkan' closely except that it matures first in January and holds its fruit in good condition until late in the spring, much like the 'Murcott'. It has never been planted on a large commercial scale, but has had a steady small demand as a home and gift-box fruit because of its high quality and long season, in spite of being somewhat seedy.

'Cleopatra' (*Citrus reshni* of Tanaka)—This handsome, small, red tangerine is far more important as a stock for other citrus fruits than as a fruit tree, although it was originally introduced for its fruit. Native to India, it had somehow reached the West Indies a century ago, and was introduced to Florida about 1875 from Jamaica as the 'Spice' tangerine by Col. C. Codrington, who had formerly lived there. The origin of the name 'Cleopatra' is entirely obscure. It occurs in horticultural literature first in 1887 in Pliny Reasoner's report, in which he says this is probably a synonym of 'Spice'. At one time Tanaka considered that 'Cleopatra' was the same as the 'Chinese Ponki', much used as a stock in China and Japan, but further study showed the 'Cleopatra' to be unknown in China and Formosa. H. J. Webber considered it "probably the most ornamental of all citrus types" because of the symmetrical tree shape, the dense, dark green foliage, and the brilliant red fruits. The fruit is of good quality, but rather tart, and is too small and seedy to be worth cultivation except as an ornamental or as a source of stock seeds. It matures in January and February.

Lemons

The true lemon, *Citrus limon* Burm., seems to have had its origin in northwestern India. It is not known in wild form and may have arisen thousands of years ago under cultivation. It reached southern Italy by A.D. 200, but may have failed to survive the Dark Ages following the overrunning of Italy by the Goths and Vandals. Tolkowsky has shown that the lemon was introduced again to Sicily before 1000 by the Moslems, if it was not still surviving there, and that it was being cultivated in Iraq and Egypt by 700. With nearly a millennium of culture in Sicily, it is not surprising that this has traditionally been the source of commercial varieties. Apparently the true lemon did not reach China until around 1200, the "lemon" mentioned earlier in Chinese literature being a related species, *C. limonia* Osbeck, from Tanaka's careful investigations. By 1100, lemons were widely known in Italy, Spain, and Portugal.

Lemon seeds were among those brought to Hispaniola by Columbus in 1493, and while they are not specifically mentioned, the importance of lemons in the Spanish diet suggests strongly that they might well have been included among the fruits listed in 1579 as flourishing at St. Augustine. Little mention is made of lemons in records of colonial Spain, but in 1839 Williams noted that lemons were increasingly being planted in northeastern Florida. Grown entirely as seedlings, lemon trees would often recover from the occasional freezes of this area.

Commercial lemon culture in Florida was begun only after 1870 and was spurred by the extensive importation of lemons to the United States from Sicily. Florida orange growers (and likewise those in California) felt that this market could profitably be supplied with home-grown fruit. They recognized, however, that the seedling types already growing in this country could not compete with the superior types coming from Italy, and so set about obtaining equally good varieties with which to supply the market. Several varieties resulted simply from planting seeds of good Sicilian lemons and selecting among the seedlings for superior quality and performance. Budwood was also imported in a few cases. An industry rapidly developed in both Florida and California, but in addition to the handicap of a freeze in 1886, the lemon suffered two additional disadvantages in Florida. In the humid climate of this state the lemon scab disease was hard to control, and proper curing of lemons was difficult. After the freeze of 1894–95, lemon culture was almost wholly abandoned as a commercial proposition in Florida, although a small lemon culture for home and local use was continued.

Since 1953 there has been a renewed interest in growing lemons in Florida and several large plantings have been made. This interest is chiefly in production of lemons for making frozen lemonade concentrate, rather than for fresh shipment to the market. This involves no need for curing the fruit, and control of lemon scab is not a real problem with the fungicides and spray equipment available today. Florida growers are confident that they can produce lemon juice at a lower cost than their competitors. However, cured Florida lemons are very well accepted by the housewife and the demand within the state is large. It is felt that it may prove profitable to make spot-pickings for the fresh market early in the season (August-September), while later fruit will go to processing plants. The tendency for the size of Florida-grown lemons to be too large for the grocery trade is no disadvantage in production for processing. The processor asks only for: (1) high juice content, (2) high acid content, and (3) absence of any unpleasant aftertaste. These qualities Florida lemons can provide. The only real problem is the somewhat greater cold hazard than for oranges, and this only limits the area suitable for lemons.

Plantings of lemons in the southern portion of the state have gradually increased to about 10,000 acres with continued interest in further development. Most of the acreage is located near Lake Okeechobee in Martin and Palm Beach counties, an area that is relatively frost-free. Smaller plantings also occur in Hillsborough County and other locations which afford the proper climatic protection. Production has exceeded 1 million boxes per year since 1969.

Lemon Varieties

There are basically only two types of commercial true lemons, and these were apparently selected long ago in Sicily and in Portugal. The various named varieties are selections within these two types, represented by 'Eureka' and 'Lisbon'. As analyzed by H. J. Webber, the 'Eureka' type is rather sturdy, characterized by an open, spreading tree habit with relatively few branches and twigs, and leaves which are dark green and rather blunt-pointed; while the 'Lisbon' type makes a rather dense tree with many slender, upright branches, and leaves which are light green and acute at the tip.

'Eureka'—This is the principal lemon variety of California but it is not grown under that name in Florida to any extent. It originated as the best seedling from seeds of a Sicilian lemon planted in California in 1858, and was introduced as a variety in 1878 by T. A. Garey of Los

Angeles as 'Garey's Eureka'. Within two years, however, the name had been shortened to 'Eureka' alone. Thorns are few, acidity is high, and seeds are few, but fruit is mostly borne on the tips of the branches, so that it is somewhat subject to damage from wind and sun. The tree is more subject to injury by cold, also, than most other varieties. The fruit mostly matures in Florida from August to December, although some fruit may mature all year long.

In California the 'Eureka' has yielded heavy crops but has tended to have a short bearing life because of susceptibility to shell-bark and dry-bark diseases. However, Dr. H. B. Frost has developed a nucellar strain which is free from these virus infections, and this strain has recently been widely planted. Many clonal selections are now being used in California.

'Lisbon'—This variety is second place in California, but it is not grown under this name in Florida now, although prior to 1895 it was considered the best lemon grown here. It had its origin in Australia from seed of lemons from Portugal and was introduced to California from Australia in 1874 as budwood. Like 'Eureka', it was first propagated commercially by Garey around 1880. Trees are much more thorny than are 'Eureka' trees but are more vigorous and productive. Acidity is as high, there are usually more seeds, and the fruit tends to be produced mostly inside the tree canopy where it is better protected from sun, wind, and cold than in 'Eureka'. Selections of outstanding trees for propagation by nurserymen in California have given rise to many clonal types.

'Sicily'—The variety planted under this name in Florida since 1953 is apparently not the Sicilian type at all but somewhat resembles the strain of 'Lisbon' known in California as Short-thorned Lisbon. The variety 'Sicily' which was imported from that country by Gen. Sanford in 1875 has apparently disappeared from Florida culture. The presently planted 'Sicily' was found about 1952 as an old seedling tree in the Bearss grove near Lutz, Florida. The parent tree is believed to have been planted about 1892. A selection was made and planted extensively near Babson Park in 1953 by Libby, McNeill and Libby. This grove was sold in the late 1960's and is no longer in production, although it supplied budwood for the company's later plantings in Palm Beach County. This variety is now referred to as the 'Bearss' lemon; large plantings have been made in Martin, Palm Beach, and Hendry counties.

'Avon'—This variety originated as a budded tree of unknown origin at Arcadia. A tree budded from this about 1934 by Mr. J. H. Jones in the Alpine grove near Avon Park, of which he was the manager, attracted the attention of Mr. W. F. Ward by its heavy production of high-quality fruit for frozen juice. The variety has been propagated since 1940 by Ward's Nursery as the 'Avon'.

'Harvey'—About 1940 the parent tree was discovered by Harvey Smith growing in the home grounds of George James in Clearwater. It was already there when he bought the property and the original source is unknown. Because the trees budded from this fruited well on several stocks and stood up well in a freeze, the Glen St. Mary Nurseries Company began commercial propagation in 1943, calling it 'Harvey'. It is apparently very similar to 'Eureka'.

'Villafranca'—Introduced to Florida in 1875 from Sicily by Gen. Sanford, this variety has been grown continuously since that time and until recently produced most of what few lemons were borne in Florida. It is very similar to 'Eureka', distinguished by greater vigor and thorniness while trees are young, but almost impossible to differentiate by fruit characters or mature tree habit. In spite of its long proven adaptation to Florida conditions, it is not being as extensively planted yet as some other varieties.

'Ponderosa'—Not a commercial variety, this has been a favorite for home planting since around 1900. It originated around 1886 as a seedling of unknown source grown by George Bowman of Hagerstown, Maryland, and was first propagated as a greenhouse plant under the name 'American Wonder' lemon. The fruits resemble ordinary lemons except for being much larger, and are grown more for the curiosity of their size than for the juice they produce; yet the juice is abundant and of good acidity, although seeds are numerous. The tree is dwarf in size and fairly thorny. This is probably not a true lemon, but if it is a hybrid, the experts do not agree on its probable parents. It is even more tender to cold than true lemons, although not more so than limes. Small-scale commercial plantings have been made since 1948.

'Meyer' (*Citrus meyeri* Tanaka)—Definitely not a true lemon and not even looking very much like a lemon, this variety is commonly called a lemon because the juice tastes much like one. It was introduced in 1908 by Frank N. Meyer for the U.S. Department of Agriculture from China, where he found it cultivated near Peking as an ornamental pot

plant. The mature fruit greatly resembles a large orange in both external and internal appearance, although it may possess a low nipple at each end, and is very juicy, with smooth, thin rind and about 10 seeds. Juice quality is good but acid content is lower than is usual in true lemon and the rind lacks the lemon aroma. Indeed, the different character of the peel oil is the biggest handicap to larger use of this lemon for concentrate. Even a small amount of Meyer peel oil gives an undesirable flavor to the juice, though this can be masked, if not too intense, by adding peel oil of true lemon. Some fruit matures at all seasons, but the main crop is from December to April. The tree is small and bushy, nearly thornless, and much more resistant to cold than any true lemon. It has been planted on a rather large commercial scale in Texas and has been commercially planted on a small scale in Florida since 1930. Because this variety usually carries the tristeza virus it should not be planted on sour orange stock or in areas where tristeza is active. It is prohibited by law in some areas of California. By reason of its cold-hardiness it is popular throughout the orange belt of Florida as a lemon tree for the home grounds, since it bears heavily and needs a space only about 8 or 10 feet square. Undoubtedly the 'Meyer' is a hybrid, but the other parent besides lemon is entirely unknown, although the color suggests orange or mandarin.

'Rough'—Used in Florida only as a stock, this lemon does not belong to the same species as the true lemon but is *Citrus jambhiri* Lush. Native to India, where it is the most important citrus stock, this lemon must have reached Florida early in the nineteenth century, but no record exists of its introduction. H. J. Webber thinks it was introduced to the West Indies in the seventeenth century, but there is no known mention of its existence there or in Florida until the last quarter of the nineteenth century. By that time it had escaped from cultivation and become widely naturalized, being known as 'Florida Rough' and as 'French'. In southern Africa it became wild along the banks of the Mazoe River in Southern Rhodesia, and so is known as 'Mazoe' lemon in the citrus industry of South Africa, where it is the principal stock. It is also the principal stock used in Florida, but while it was already so used in 1876, it was not until the extensive planting of the deep sandy soils of central Florida that this usage became dominant. With the appearance of young tree decline (YTD) in Florida, especially on trees budded to 'Rough' lemon, this stock has been severely restricted in new plantings. Only a very small amount is currently (1973) being used by nurserymen. The tree is very thorny and the fruit is much flattened, with very thick, bumpy rind. The juice is much like

that of true lemon, though lower in acidity, and the mature fruit is lemon-colored.

Limes

The true lime, *Citrus aurantifolia* Swingle, is a small, thin-skinned, very acid fruit, native to the Malaysian area. It does not occur in China at all, being too tender to cold, but is found in many forms in eastern India, and perhaps is truly wild in northern Malaya. It is not certain whether the lack of mention of this species in early citrus literature is because it was confused with the lemon or because it was less hardy to cold. At any rate, it was not until the thirteenth century that we find the lime recorded in Italy, and the seventeenth century before it received more than casual mention in Europe.

While Las Casas does not specifically include the lime among the seeds brought by Columbus to Haiti in 1493, he does not pretend to give an exhaustive list. Since Oviedo reported limes as plentiful there in 1520, it is quite probable that Columbus brought these seeds also from the Canaries. The lime soon became naturalized on some of the West Indies, on the coast of Mexico, and ultimately on some of the Florida Keys. The first mention of it in Florida is by Williams in 1839, who noted that planting of limes was increasing. In 1838 Dr. Henry Perrine planted a few lime trees from Yucatán on Indian Key and perhaps some adjacent islands, and naturalized lime trees found on many of the Keys at the turn of the century have often been considered as having resulted from his plantings. However, in 1876 his son was unable to find any trace of those plantings still in existence except for the thriving sisal plants and a few palms. Throughout the nineteenth century the common lime, variously known as 'Key', 'Mexican', or 'West Indian', was primarily a home fruit in Florida, although there was some small commercial culture by 1883 in Orange and Lake counties.

After the culture of pineapples was abandoned on the Keys in 1906, as the combined result of depletion of organic matter and the 1906 hurricane, there developed slowly a lime industry in its place. Plantings increased rapidly after 1913 on the Keys and on islands near Fort Myers, with production reaching a peak in 1923. The hurricane of 1926 gave this industry a reverse from which it never recovered.

Meanwhile the so-called 'Tahiti' or 'Persian' lime had entered the picture. This is not a true lime, being probably a hybrid of some sort, and is called *Citrus latifolia* by Tanaka. Undoubtedly it is a hybrid resulting under cultivation, for it is known only as a garden plant appearing in California about 1875. From 1850 to 1880 oranges and limes were imported in great numbers to San Francisco from Tahiti, and seeds of these fruits were sometimes planted. The parent tree of the Tahiti type of lime must have arisen from such a seed, representing a chance cross which occurred in Tahiti. The variety is triploid in chromosome number, a condition usually producing sterility. No one seems to have recorded anything about this tree, and our first record is the statement by Garey in 1882 that the 'Tahiti' lime is worthless in California. The next year Rooks noted that the 'Tahiti' lime was growing in Florida, and that he knew by report of the 'Persian' lime. For many years 'Tahiti' and 'Persian' appeared as distinct types, with one sometimes rated inferior to the other, but the two names have long been considered synonymous. No explanation has ever been offered for the origin of the name 'Persian' for this type of lime, which is now known in Florida chiefly by this name and is called 'Bearss' in California.

Reasoner in 1887 reported that the 'Tahiti' lime was planted in grove form in "the lake country," but the lemon was considered a more promising commercial fruit of acid type and plantings of 'Tahiti' lime increased slowly. Along with the decline of 'Key' lime plantings in the late 1920's, however, much interest developed in groves of 'Tahiti' lime in the southern orange belt. Since 1930 Florida limes have been almost wholly of the 'Persian' ('Tahiti') type, with Dade County far in the lead and the southern end of the Ridge (Polk and Highlands counties) as a secondary area.

'Key' limes were so much smaller and seedier than lemons that there was never any confusion between them in the market. 'Persian' limes are seedless, as 'Eureka' lemons tend to be, and are of lemon size, so that confusion is easy. The lime grower has preferred to put his fruit on the market green as an easy means of separating it in the mind of the buyer from lemons. Like lemons, 'Persian' limes are largely picked by size. Very few 'Persian' limes are grown in California, because of the cooler climate, and so this lime is almost wholly a Florida industry. The volume of Florida limes is far smaller, however, than the volume of California lemons. 'Persian' limes, like common limes, mature at all seasons of the year, but the peak of production is during the summer months.

Most of the limes grown in the world are of the common, true lime type, and these are grown in far larger amounts in other countries than are 'Persian' limes in Florida. Mexico leads the world in lime production, with Egypt second, and the United States a poor third. Florida produces about 95 per cent of the limes grown in this country. India probably has a lime production equal to or larger than Egypt's, but several species may be involved.

Florida limes are largely shipped to the fresh-fruit market, but since 1949 there has been a considerable amount of the crop processed as frozen concentrated limeade, the volume thus used varying from one-fourth to over one-half of the crop.

Three other citrus fruits are called limes or resemble them greatly—the sweet lime, the calamondin, and the 'Rangpur' lime. None of these is a true lime and none is of any commercial importance in the United States. The sweet lime, often called sweet lemon, is widely but not commercially grown around the shores of the Mediterranean and is extensively used as a stock for oranges in Israel. It has been tested as a stock in Florida but abandoned because of its great intolerance to xyloporosis, a virus disease. However, there is some current interest in the sweet lime as a rootstock with budwood free of damaging viruses. The 'Palestine' sweet lime, *Citrus limettioides* Tanaka, is very common in India. It is not the same as the 'Mediterranean' sweet lime, *C. limetta* Tanaka, of Spain and Italy. This latter species was responsible for British sailing ships of the eighteenth century being known as "lime-juicers" and British sailors as "limeys." The lime juice carried to prevent scurvy was not the acid 'West Indian' lime but the sweet lime with a higher content of ascorbic acid.

The calamondin, *Citrus madurensis* Lour., is a small seedy lime with thin, red rind and soft, juicy, red flesh. Native to the Philippines, it was named *Citrus mitis* by Blanco, but Loureiro had seen and named it earlier in Madura, an island near Java. It is sometimes given as *C. microcarpa* Bunge, but this is a different fruit, the musk lime. The calamondin was introduced to Florida as an "acid orange" in 1899 by Lathrop and Fairchild from Panama, and was early called "Panama orange." It had come to Panama from Chile, and to Chile from China, where it has long been cultivated both as an ornamental and as a stock for mandarins. In 1908 and 1909 it was offered by Glen St. Mary Nurseries as the "To (sour) kumquat," and then not listed again until 1921 when the Philippine name "calamondin" was used.

Hardier to cold than any citrus species except the kumquats and trifoliate-orange, the calamondin is popular as an attractive home fruit to provide a lime substitute. The tree is dwarf and bushy, and is quite showy for months when the mature fruit is hanging on it, being everbearing with the heaviest crop in winter. The acid content is high, the flavor is distinctive, and juice color is attractive. Seed propagation is used, as no seedling variation has been observed. Various hypotheses have been put forward as to the possibility of the calamondin being a hybrid of lime and mandarin or lime and kumquat, but these are highly speculative. It is a species which has made its own way for untold centuries.

The 'Rangpur' lime and its near relative the 'Kusaie' lime are considered by Tanaka as lemon relatives and belong to *Citrus limonia* Osbeck. They are always called limes in the United States and are used like true limes and calamondins. They have thin rinds easily separated from the soft juicy flesh, the rind and flesh both being red in 'Rangpur' and yellow in 'Kusaie', and are often thought to represent hybrids of mandarin with lime or lemon. 'Rangpur' was introduced to Florida as seed from India in 1887 by Reasoner Brothers, and matures fruit in the fall and winter. It has been sparingly grown as a yard tree for ornament and acid juice, but has been of more interest in recent years as a stock species for possible commercial use. 'Kusaie' lime is not known to be cultivated in Florida, though popular in Hawaii.

Citrons

The citron, *Citrus medica,* has already been discussed as the citrus fruit first known to Europeans, and has been grown in the warmest areas around the Mediterranean for at least 2,000 years. Commercial culture today is mostly limited to Sicily, southern Italy, Corsica, Crete, and a few small Greek islands in the Cyclades. The pulp and juice are useless, but the thick rind is preserved in brine for use in making candied peel for fruitcakes and confections. There is a small inedible type of watermelon which is also called citron because its rind is used similarly.

It was among the citrus species brought to the New World by Columbus, but has never had extensive culture here. It was probably introduced to Florida when St. Augustine was settled, but also probably did not long survive winters there. There is no commercial production in Florida today, although a few specimen trees may be found in collections of citrus fruits. No variety is grown as such, although the common large-fruited form grown commercially in Italy is similar to what has long been seen in Florida. It is propagated readily by cuttings, or may be budded. A few seedlings may be seen of the small-fruited form with persistent styles called 'Etrog', which is used in the Jewish ceremonies of the Feast of Tabernacles.

Citrus *Hybrids*

All species of *Citrus* will cross with one another, and even with species of the closely related genera *Fortunella* and *Poncirus.* Undoubtedly such crossing had a part in the development of some of our present species in the far past, for that is a normal method of origin of new species, but nobody can say with certainty which species are primitive and which are derivative. There are several cultivated forms, such as the 'Temple' and 'Murcott' "orange," the 'Meyer' lemon, and the 'Persian' lime—none of them known to exist outside of gardens—which are universally conceded to have arisen from crossing of established species, although there is not such uniform agreement on the exact species involved. Unless exactly the same form is produced by a known cross made by man, opinions as to the crosses which produced these forms must remain speculative.

Many hybrids have been produced by man through controlled (guarded) crosses in which both parents are known with certainty. The first hybrids of this kind were produced by W. T. Swingle for the U.S. Department of Agriculture in 1897, and most of the numerous citrus hybrids which have been knowingly produced were the work of Swingle and his associates in the succeeding fifteen years. No one can predict the fruit characters which will result from any given cross, and each seed in a single fruit produced by hand-pollination may develop into a seedling with fruit and tree characters different from the others. This is true, however, only if a true hybrid embryo matures. As stated previously, the only embryos which mature in a high percentage of citrus seeds are nucellar ones, which are derived wholly from the mother plant and reproduce it exactly. In some crosses, the hybridizer may never be able to produce a hybrid, or must make dozens of crosses and grow the seedlings to fruiting before he obtains a single hybrid. Thus, it has never been possible to produce a hybrid using 'Dancy' tangerine as seed parent. All resulting seedlings are 'Dancy', regardless of pollen parent. In other crosses a large percentage of seedlings will be hybrids. Once a hybrid has been obtained, however, it can be maintained permanently as a variety by budding.

Hybrids may show characters intermediate to those of their parents, some characters of each parent, new characters which are not exhibited by either parent, or characters once present in an ancestral form. Only those hybrids which represent definite improvements over the parents in some quality should be propagated as varieties. Some such hybrids have been developed and others are possible. Qualities to be considered are flavor and taste, disease resistance, vigor, productivity, adaptation to climate, and ease of handling and shipping. Some hybrids have merit as stocks because of superior adaptation to soil conditions or climate, resistance to soil diseases or insects, and ability to enable the scion variety to produce fruit of high quality.

Citrus hybrids are classified as:

1. Intervarietal (intraspecific)—crosses made between varieties of the same species. These are not properly called hybrids and usually result in little or no more variation in seedling characters than self-pollination gives. They are of no importance in production of new varieties.

2. Interspecific (intrageneric)—crosses made between species of the same genus. These are easily produced, except for the possible small number of hybrids which may be found in the resulting seedlings when they bear fruit, and the most interesting and valuable citrus hybrids have arisen in this way, notably tangelos and tangors.

3. Intergeneric—crosses made between different genera, which must usually be very closely related and must always be within the same family. Intergeneric hybrids are more difficult to obtain than interspecific hybrids, and many attempts have resulted in failures. When the crosses are successful, vigorous hybrids have resulted which are usually sterile, that is, produce only nucellar embryos. Some promising stock varieties are intergeneric hybrids, the citranges.

4. Complex—crosses made between interspecific or intergeneric hybrids and other species or genus, thus the result of two or more crosses. The difficulty of making such hybrids is even greater than for simple bigeneric hybrids, and only recently have any commercially promising forms arisen in this way.

Citrus hybrids are known by class names, equivalent to "orange," "lime," etc., and such class names are sometimes tongue twisters. In practically all instances they were coined by W. T. Swingle to identify the types of hybrids he had developed and consist of portions of the name of each parent, though sometimes these are difficult to recognize. Class names will be explained as they are used. Varieties of hybrids, like varieties within a species, are individual seedlings (or sports) propagated by vegetative means.

Classes and Varieties of Hybrids

Tangelos—These are interspecific hybrids resulting from crossing tangerines with grapefruit (*C. tangerina* x *C. paradisi*). When the name was coined in 1905, horticulturists were trying to establish "pomelo" as the accepted name for grapefruit, so that *tang*(erine) and (pom)*elo* were combined into *tangelo* to indicate this cross. The first tangelos, 'Sampson' and 'Thornton', were named and released in 1905, and were quickly planted on a small commercial scale. They did not prove very successful and are little planted today. In 1931 'Lake', 'Minneola', and

'Seminole' tangelos were released and have proven much more satisfactory, so that they are being planted in increasing numbers. Certain marketing groups have worked diligently to maintain quality in shipments and to stimulate market demand. The Tangelo Act of 1955 placed these varieties under maturity regulations for the first time and was a further aid to orderly marketing of fruit of good quality.

'Orlando' (*Lake*)—The result of 'Dancy' tangerine pollen on 'Duncan' grapefruit pistil, a cross made by W. T. Swingle in 1911. It has little resemblance to either parent and has come to be accepted by the market as an orange under the name "Orlando orange," which has replaced its original name 'Lake' tangelo. Size and shape are tangerine-like, but the color and texture are orange-like. Rind color is deep orange, as is the pulp color, and the rind adheres firmly to the pulp. It has about 10 seeds and matures in November and December. It has been planted on a large commercial scale, totaling some 20,000 acres in 1970, because of its sweetness and juiciness in early winter. 'Temple' or 'Dancy' trees should be interplanted with 'Orlando' for satisfactory crops, as it requires cross-pollination.

'Minneola'—This variety resulted from the same cross as 'Lake', but shows much more evidence of its tangerine parentage, although it also does not have the loose-skin character of the mandarins. 'Minneola' has a delicious flavor and aroma when fully mature, is strikingly handsome with dark reddish-orange rind and dark orange flesh, and is also very juicy. It has been handicapped somewhat by low productivity, but this is now known to be due to need for cross-pollination and should not be a commercial problem. 'Lee' and 'Robinson' have been suggested as pollinators. It is also subject to attack by citrus scab, and the fruit shape does not lend itself to easy packing. Seeds average about 10, and the season is late, January to March. There were about 1,000 acres in 1970 in small blocks, grown especially for the local retail and gift-box trades.

'Seminole'—The third commerical variety from the cross producing 'Lake' and 'Minneola', this one has been planted much more sparingly than 'Minneola', which it resembles greatly. It differs from 'Minneola' chiefly in having a more easily peeled rind, less susceptibility to scab, a much larger number of seeds (22), and a later season of maturity (February to April). Flavor, aroma, and juiciness are good, though the taste is a trifle tart usually.

'Sampson'—The result of using 'Dancy' pollen on an unnamed grapefruit variety in 1897 by W. T. Swingle; this was the first tangelo developed. Like the above tangelos it is very juicy, highly colored, and moderately seedy (about 15 seeds), but it is quite tart and is best suited to making an "ade." There is also a trace of the bitterness characteristic of grapefruit but absent in the above tangelos. The combination of high acidity, very thin rind, and easy bruising, together with high susceptibility to scab, made 'Sampson' unsatisfactory commercially. Its season of maturity is February to April.

'Thornton'—This variety resulted from the same type of cross as 'Sampson' but was made two years later. It differs markedly from all the above tangelos in having a thick rind, easily separated from the flesh, which becomes quite loose when overmature, making puffy fruit. Rind and pulp are both light orange in color and seeds number about 15. The pulp is soft, juicy, and sweet, being insipidly so when overmature. The season of maturity is December to February. 'Thornton' has never been planted on a large commercial scale, though it is popular with some specialty growers.

Tangors—These are interspecific hybrids between some type of mandarin and the sweet orange, the name being a combination of *tang*(erine) and *or*(ange). Of the tangors produced by known crossing only the 'Murcott' has attained any commercial importance, but the creation of these hybrids has made it practically certain that some hybrids which have occurred by chance under cultivation are natural tangors. Notable among these is the 'Temple', which is an important commercial variety, usually sold as an orange. The 'Umatilla' tangor is a known hybrid that has been planted to a small extent. Some authorities believe the 'King' mandarin is a tangor, but if so it is a very ancient one which has persisted for many centuries on its own, whereas all of the above have arisen in very recent times in gardens. In general the tangors have flesh texture and flavor more like oranges than mandarins, but have rinds separating fairly readily from the flesh, though less so than do mandarins. Sugar content of juice and juice content of tangors are often greater than is usual with oranges, but fruit cannot be held on the tree as many months after maturity as can oranges.

'Temple'—By far the best known and most important tangor is marketed as "Temple orange." It had its origin in Jamaica sometime late in the nineteenth century, but first came to horticultural attention in

even a small way about 1896. A fruit buyer named Boyce, who usually obtained his supply for northern markets in the Oviedo area, went to Jamaica for oranges following the big freeze of 1894–95 because there were none in Florida. Here he ran across a seedling orange which pleased him greatly, and sent budwood to friends at Oviedo—Rev. R. W. Lawton, and Messrs. J. H. King and J. H. Lee. All of these budded a few trees with the 'Jamaica' orange. A close friend of Mr. King, Allan Moseley, was caretaker of several groves around Winter Park at this time and is believed to have obtained budwood from Mr. King when his trees fruited to bud a tree in the young grove of Mr. Wyeth, of which he was caretaker, at Winter Park, about 1900. This grove was bought by John S. Hakes in 1910, and sold by him in 1914 to his son Louis A. Hakes. At any rate, in 1915 one tree in the Hakes grove was brought by L. A. Hakes to the attention of his neighbor, W. C. Temple, formerly manager of the Florida Citrus Exchange; and Mr. Temple, in turn, called the attention of M. E. Gillett, his associate in the Exchange and one of the leading citrus nurserymen, to the unusual qualities of the new variety. After testing it out on several stocks, Mr. Gillett signed a contract with Mr. Hakes in February, 1916, for the exclusive propagation of this variety, not yet named. It was Mr. Temple's desire to have it called "Winter Park Hybrid," although editor Edgar Wright of the *Florida Grower* urged that it be named 'Temple'. After propagating a large supply of nursery trees, Buckeye Nurseries, of which Mr. Gillett was president, first announced this variety publicly in May, 1919, under the name 'Temple' as a posthumous honoring of W. C. Temple. Trees were expensive and were sold only under contract not to allow any propagation from them. In 1921 Messrs. W. J. Lawton and C. S. Lee, sons of the original Oviedo planters, went to see the Hakes tree and informed Mr. Gillett that the 'Temple' had been growing in Oviedo for 25 years. Thereupon the rights to propagation of all of those trees in Oviedo were bought by the Glen St. Mary Nurseries Company, which had bought out Buckeye Nurseries, thus protecting the name 'Temple'.

The 'Temple' was not the overwhelming success which had been anticipated, in part because of propagation in considerable measure on 'Rough' lemon stock, under inadequate mineral nutrition. The fruit proved somewhat hard to ship when given the same packinghouse treatment as oranges. The quality is exceptionally fine with proper fertilization, however, and well-grown fruit usually commands a good premium over oranges. The season of maturity is January to March, or sometimes April. Planting of 'Temple' has been heavy since 1940, and in 1970 there were more than 20,000 acres of this variety.

'Temple' fruit is highly colored and handsome, a deep orange or reddish orange with about 20 seeds. It is somewhat subject to citrus scab. Quality of flesh is slightly better on 'Cleopatra' than on other commonly used stocks, but color is not so high. Good quality is obtained on all of the standard stocks with good fertilizer programs.

'Murcott'—This old variety has been planted extensively only since 1952. The parent tree was sent for trial from the U.S.D.A. nursery at Little River (Miami), Florida, to R. D. Hoyt of Safety Harbor about 1913. Undoubtedly it was a tangor from the breeding work being carried out by Swingle and his associates at that time, but the identification label was lost. A neighbor, Charles Murcott Smith, obtained a bud of this tree from Mr. Hoyt and propagated a few trees about 1922. Small-scale commercial propagation was undertaken in 1928 by the Indian Rocks Nursery under the name 'Honey Murcott', and several other nurseries on the West Coast propagated the variety in a limited way prior to 1950. In 1944 J. Ward Smith (no relation to C. M.) became interested in commercial production and made near Brooksville the first planting for commercial shipment. He used the name 'Smith Tangerine', apparently unaware that the name 'Murcott' had previously been approved by C. M. Smith. The spellings 'Mercott' and 'Marcott' have been used but are incorrect.

As no technical description of the 'Murcott' has been published previously, one is given here.

Fruit color yellow-orange (R.H.S. Orpiment orange 10/1); surface smooth, glossy, shallowly ribbed to conform with segments; shape oblate; size medium, diam. 7.0 to 8.0 cm., height 4.7 to 5.2 cm.; base almost flat, or with very shallow depression 2 cm. across and 1 mm. deep; apex same as base; calyx small, lobes 1 mm. long; rind thin, about 1.2 mm., tightly appressed to flesh but peeling fairly easily until fruit has been off the tree some days; oil glands spherical, deeper color than rind, small, surface level; axis semihollow, elliptical, 15 mm. x 5 mm.; segments 11 to 12, membranes thin but tough; pulp orange-colored (R.H.S. Marigold orange 11), tender but not melting, very juicy, with little rag; vesicles medium size, fusiform; flavor rich, sprightly; seeds 18–24, small, pointed and rounded, 10 mm. x 6 mm.; embryos white; season January to March.

'Murcott' is somewhat subject to scab and is reported somewhat subject also to xyloporosis. The tree is bushy, with willowy branches, and fruit is sometimes wind-scarred because it is borne mostly near the ends of the branches. This position also makes fruit more subject to frost injury. Like 'Temple', the 'Murcott' runs higher in soluble solids than sweet oranges of comparable maturity. The trees tend to bear heavily. Since 1952 the variety has been planted quite heavily each year, with 10,000 acres reported in 1970.

'Umatilla'—This variety came from pollination of 'Ruby' orange by 'Owari' satsuma in 1911 by W. T. Swingle, but was not offered for propagation until 1931. The fruit is the size and shape of a 'King' mandarin and matures about the same time, although the rind is redder orange than 'King' and quite smooth. Quality is excellent but the variety has never been planted on any but a very small scale. It is a fine late variety for home use and gift-box shippers. Because of its general similarity to the tangelos it is sometimes known, incorrectly, as 'Umatilla' tangelo.

Orangelos—This is a group of varieties which have the general appearance of the grapefruit without its bitterness and aroma and are considered to be natural interspecific hybrids of *Citrus paradisi* x *C. sinensis.* The 'Triumph', discussed under grapefruit as the first named variety, along with the 'Imperial', 'Royal', and 'Chironja' should probably be so classified.

'Chironja'—This variety was discovered in the mountains of Puerto Rico and described by Moscoso in 1956. It approximates an average grapefruit in size with orange pulp and bright, glossy yellow rind at maturity. It is a juicy, sweet fruit which can be juiced, halved and eaten with a spoon, peeled and eaten by sections similar to a tangerine, or processed for canning. The rind can be utilized in marmalades or candied peel. It is used as a shade tree for coffee, coming true to type, undoubtedly nucellar, from seed. The name is derived from "China" (sweet orange) and toronja (grapefruit) of the Spanish.

Limequats—These are intergeneric hybrids, produced by pollinating 'Key' limes (*Citrus aurantifolia*) by kumquats (*Fortunella* spp.). Two of these hybrids, from crosses made by Swingle in 1909, have been sparingly grown in Florida. It was hoped that they would extend lime culture farther north by adding some of the cold-hardiness of the kumquat to the usual fruit qualities of the lime. This hope has been realized to some degree, but the hybrids are about like 'Temples' in hardiness and not quite equal to sweet oranges. They are very resistant to the withertip disease which has seriously limited growing of common limes on the mainland, and they are quite prolific producers of a fruit which closely resembles limes. Yet they have never become very popular and are planted solely as home fruits.

'Eustis'—The result of 'Key' lime pollinated by 'Marumi' kumquat. The fruit is nearly oval, about 1¾ by 1¼ inches, light yellow at maturity, with around 8 seeds. The main crop is borne in late fall and winter, but there is some at any season. The distinctive feature from the lime parent, besides greater hardiness to cold and disease, is the sweetness of the rind instead of the bitter flavor of lime rind. Trees have abundant small thorns.

'Lakeland'—This variety was developed from another seed of the same fruit as 'Eustis'. It has much larger fruit than 'Eustis', 2¼ by 1¾ inches, and is of a little deeper yellow color, but similar in shape and acidity. Seeds are fewer (about 5) and larger, and the twigs are nearly thornless. Some years ago there was a small vogue for planting this variety as "Lakeland golden lime," but it did not last long.

Citranges—Webber and Swingle described the first crosses between *Poncirus trifoliata* and *Citrus sinensis,* naming them 'citranges', in 1904. Their desire was to impart the cold-hardiness of the trifoliate-orange to a hybrid with the delectable qualities of the sweet orange. The fruits fail to come up to the standard of quality of the sweet orange although they retain some of the cold-hardiness of the trifoliate parent. The 'Rusk' and 'Troyer' citranges have been given consideration as rootstock varieties, along with other named varieties. There is currently a move in Florida to use the 'Carrizo' citrange as a rootstock to overcome the difficulties besetting some of the more common rootstocks which have had such long usage in the state. Some evidence indicates that the 'Troyer' and the 'Carrizo' are of a single clone which resulted from the same cross ('Washington' navel orange x trifoliate-orange), made under the direction of Dr. W. T. Swingle in 1909.

Citrumelos—These intergeneric hybrids resulted from crosses between *Poncirus trifoliata* and *Citrus paradisi.* They are very similar to citranges and, like them, hold the principal interest as possible rootstocks for other types of citrus.

Citrangequats and Citrangedins—These are examples of the numerous classes of complex hybrids created by Swingle and his associates, none of which has any commercial importance. Citrangequats represent crosses of the kumquat (*Fortunella* sp.) with the hybrid citrange and are trigeneric crosses. Citrangedins result from crossing citrange with calamondin (*Citrus madurensis*) and so are bigeneric only, although complex. The 'Thomasville' citrangequat and the 'Glen' citrangedin have both been described and propagated in a very limited way for use as lime substitutes in areas like Georgia which are too cold even for the calamondin.

In 1959, three new early maturing tangerine-tangelo hybrids, ('Robinson', 'Osceola', and 'Lee') were introduced by the U.S.D.A. Agricultural Research Service from its Orlando station. They were selected from crosses of 'Clementine' ('Algerian') tangerine and 'Orlando' tangelo, being therefore complex in nature. 'Robinson' and 'Osceola' closely resemble tangerines, while 'Lee' has characters resembling more the sweet oranges or tangelos. Excelling 'Dancy' tangerine in size, sweetness, and earliness, it is anticipated that they will find an important place in the fresh-fruit trade, particularly as specialty fruits. 'Robinson' is the earliest of the three, maturing in late October. It bears regularly and has 10–20 seeds. 'Lee', which resembles 'Orlando' (its pollen parent) in size and shape, develops good rind color and eating quality in late October or early November. 'Osceola' is at its best in November and has unusually high rind color, but is more tart than the other two and a little seedier (15–25 seeds).

The 'Page' orange, resulting from a cross between 'Minneola' tangelo and 'Clementine' ('Algerian') tangerine, was introduced by the U.S.D.A. in 1963. The bright orange fruit has an excellent flavor, ripens in October, and is at its prime in November. It can be peeled easily but in mixed plantings contains from 10 to 25 seeds and tends to be small. The 'Nova' tangelo, introduced in 1964 by the U.S.D.A., is a hybrid of 'Clementine' tangerine and 'Orlando' tangelo, the same cross which produced the 'Robinson', 'Osceola', and 'Lee' tangerines. In Florida, the fruit, orange to red with numerous seeds in mixed plantings, resembles early-ripening 'Orlando' tangelos in size and shape although the trees have distinctive mandarin characteristics.

These five U.S.D.A. hybrids appear to set more fruit and develop higher seed numbers with cross-pollination, characteristics of the 'Orlando' and 'Minneola' parents. The 'Robinson' has been planted quite widely since its introduction and has attained good market acceptance. Following its introduction, the 'Nova' quickly attracted the attention of the industry and currently appears to be leading the other four hybrids in propagation. The 'Lee' and the 'Osceola' have not had widespread acceptance while the small size and possible scab susceptibility of the 'Page' rule against its general commercial acceptance.

3. Propagation of Citrus Fruits

THE first citrus trees grown in Florida were seedlings, since seeds were far easier to transport on long sailing voyages than budded trees, and for two and a half centuries thereafter it seems unlikely that any method of propagation other than by seed was practiced here. Budding was standard practice in Europe during the sixteenth and seventeenth centuries, if Galesio was well informed, but during the eighteenth century there came a realization that seedling orange trees were larger and more productive, and budding went out of fashion in Spain and Italy. This was undoubtedly largely responsible for the delay in budding citrus trees in Florida.

Historical Background

The first budding of which we have record was in developing the Dummitt grove on the upper end of Merritt Island in 1830. Zephaniah Kingsley had imported budded orange trees from Spain in 1824 for planting what was later known as the Mays grove at Orange Mills, a few miles north of Palatka. Buds from this grove were taken for budding the wild sour orange trees of the Dummitt grove, according to E. H. Hart in 1877. Thereafter we read of occasional groves being budded, but far more were planted by seed. The phenomenon of nucellar embryony in citrus seeds was another potent factor for delaying recognition of the superiority of budded trees. Seedling apple trees bear fruit almost always much inferior to the parent tree, but a high percentage of orange seedlings come from nucellar embryos which reproduce the parent exactly. There are always some variant trees in a seedling grove, however, and all trees are usually more thorny when grown from seed and require much longer to come into commercial production.

In the 1870's there began to be really large-scale development of orange groves around Orange Lake and Lake Weir by budding wild sour orange trees, and for this purpose budwood was carefully selected from a very few trees. Citrus growers began to be interested in clonal* varieties derived from one parent, rather than in seedling types, and budded nursery trees were imported of various named varieties from Europe. Mostly these came from the Rivers Nursery in England, but some were brought by Gen. H. S. Sanford from the Mediterranean area. In 1877 the first short list of such varieties with descriptions was drawn up by a committee for the Florida Fruit Growers Association. Thereafter the varietal list grew apace, and more and more groves tended to be budded trees rather than seedlings, although many growers continued to argue the merits of the seedling grove. Around 1880 several treatises on orange culture appeared in both Florida and California, all stressing the superiority of budded trees, and thereafter few seedling groves were planted.

Any method of propagation other than by seed will assure the advantages of early bearing, few or no thorns, and uniform fruit season and quality that distinguish the budded tree from the seedling. While stem cuttings or layers may be used for these ends, only budding or grafting can give the additional advantages of having a root system which is better adapted to the soil conditions than the scion roots are, or which can confer greater hardiness to cold on the aerial part of the tree. Grafting was long ago found much less satisfactory than budding for propagating citrus trees, except for topworking old trees in some situations, and budding is universally employed for all kinds of citrus fruit except citrons, which are often grown from cuttings. Approximately 95 per cent of all Florida citrus trees are budded, the great majority of seedling trees being quite old, remnants of early plantings.

*Clone, or clonal variety, is a term introduced by the late H. J. Webber to denote plants derived by asexual (vegetative) propagation from a single original plant. Thus, all trees of the clone 'Pineapple' sweet orange may properly be considered as branches of the original 'Pineapple' tree. In recent years the term "cultivar" (abbreviated "cv."), for "cultivated variety" has been promoted in horticultural literature as an international term to distinguish horticultural from botanical varieties. All clones are cultivars, but by no means are all cultivars clones. It is quite correct to write "sweet orange cv. Pineapple" or "variety Pineapple," but the dominant usage is to write "sweet orange 'Pineapple'," the single quotes indicating that the name is that of a variety (cultivar).

Stocks

Stocks for budding citrus varieties are practically always grown from seed today. In the case of a variety which produces few seeds, such as the 'Rusk' citrange, cuttings may serve in place of seedlings. Thanks to the high percentage of seedlings of any of the common stocks which develop from nucellar embryos, there is a high degree of uniformity in a bed of seedling stocks. The few seedlings which show variation (usually less vigorous) can be rogued out easily. Several different species or hybrids have been used as stocks, each having some merit under certain conditions. Commercially only a small number have proved desirable. A satisfactory stock must be congenial with the top budded on it; that is, the two must form a union which permits good growth, long life, good yields, and good fruit qualities of the scion variety. Any dwarfing effect is an indication of a certain degree of uncongeniality, or of the presence of a systemic disease, although yield may be good for the tree size and fruit quality excellent.

Besides congeniality with the scion, the stock must possess certain abilities to adapt itself to the various factors of its environment if it is to be successful. Among these are: (1) *Nursery adaptability*—ease of handling from the nurseryman's standpoint. This includes ready availability of seed, high percentage of nucellar embryos, vigorous seedling growth, relative freedom from pest attacks, and easy budding. (2) *Soil adaptability*—the relative vigor of growth on soils of varied depth, texture, structure, reaction, salinity, moisture, and nutrient supply. (3) *Climatic adaptability*—the degree of hardiness to cold conferred by the stock. (4) *Biotic adaptability*—the degree of freedom from, or resistance to, various soil pests and diseases, or the effect of the stock—in its relation to the scion—on the resistance of the system to various disease complexes.

Some stocks are superior in one or more of these qualities, but inferior in others, and none is outstandingly superior on all counts. The same stock may be superior for one scion variety under given environmental conditions and inferior for another variety, although usually all varieties of a species will give similar responses on the same stock in the same environment. The choice of a stock, therefore, must be made for a particular stock-scion combination in a particular environmental condition. All of the stocks commonly used in Florida have satisfactory nursery adaptability. The six stocks discussed below have been used by the Florida industry over long periods of time. The researchers and commercial growers have been able to determine, with a good deal of assurance, the responses which may be obtained from these stocks with various scion varieties and under differing environmental conditions. Three of them, 'Rough' lemon, sour orange, and 'Cleopatra' tangerine, account for 90 per cent of the current plantings in the state.

1. 'ROUGH' LEMON (*C. jambhiri*)—this stock is especially suited to the deep, well-drained, sandy soils which are subject to marked fluctuations in soil moisture. On these soils it produces larger trees in shorter time than any other stock. Its use is restricted generally to areas south of a line running from Leesburg to Sanford by its tenderness to cold, and to properly drained soils by its susceptibility to foot rot. It is tolerant of the tristeza and exocortis viruses, but can be affected by xyloporosis. It forms a normal union with all scion varieties, develops a deep root system, produces high yields of fruit, and permits long life of the tree when planted on suitable soil. Its only real disadvantage in most cases is that fruit quality is a little lower on this stock than on the other common ones. Often the higher yield more than compensates for this effect, but not when the scion is a satsuma or one of the tangerine-type hybrids such as a tangor or tangelo.

Prior to 1865, apparently only sour orange and sweet orange were used in Florida as stocks. Soon thereafter the "native" 'Rough' lemon attracted attention as a stock giving faster growth and earlier bearing for oranges and lemons. As the high pineland soils were planted more extensively in the southward development of the citrus industry, 'Rough' lemon more and more replaced older stocks for use on these soils, although the recommendation of it by Pliny Reasoner in 1887 for low, wet land did not prove sound.

'Rough' lemon was the rootstock used on about 70 per cent of the citrus acreage of Florida in groves planted from the 1920's to the 1960's. However, in 1964, at Wauchula (Hardee County), a six-year-old grove of 'Pineapple' orange on this stock began to show symptoms of a decline which was subsequently called "young tree decline." During the intervening years this decline (discussed more fully in Chapter 7) weakened many young trees in the flatwood and marsh areas; then, a similar type of decline, called "sand hill decline," was noted in the well-drained areas in the southern portion of the citrus belt. Because this decline appears most prevalent on 'Rough' lemon budded to sweet oranges, particularly of the 'Valencia' variety, it has also been called "lemon root decline." Although further observations appear to indicate that other stocks may also be affected, after losses of more than 80 per cent in some groves on 'Rough' lemon, growers are reluctant to continue the use of this stock. This decline has been a serious blow to the citrus industry, and no solution is, at present, available.

The gravity of this problem is evidenced by the fact that citrus nurseries have reduced propagations on 'Rough' lemon stock from 60 per cent of total nursery inventories in 1960 to less than 10 per cent in 1973.

2. SOUR ORANGE (*Citrus aurantium*)—this species is the preferred stock for the low hammock and flatwood soils with fine texture and high water table, where its resistance to foot rot and other gumming diseases gives it real advantage. It is sufficiently hardy for use in all sections of the state where oranges are grown commercially. Trees on sour orange are of standard size, in spite of a slight tendency for the scion to overgrow the stock, and produce good crops of excellent quality. The sour orange is, unfortunately, very much affected by the presence of the virus disease, tristeza, with almost all scions except lemons. The citrus industry of Brazil, with nearly all trees on sour orange, was practically wiped out in the 1940's by a severe form of this virus. The disease is present in Florida in much less virulent form and few groves on sour orange have been eliminated although some injury has occurred particularly with sweet orange scions. There is, however, evidence of some spread of the disease by aphid vectors, varying in amount from season to season. The disease holds a potential threat to the industry. In spite of this, and excepting a few years after tristeza was found in the state in 1951, sour orange has been held in favor by many growers. In recent years it has been used to the extent of 40 per cent of nursery propagations, gaining somewhat in popularity with the discovery of young tree decline on 'Rough' lemon stock. While approximately 25 per cent of the citrus trees in the state are on sour orange stock, it should be used with extreme caution, and its current popular use as a rootstock for sweet oranges and mandarins is discouraged. ? NA

It should be noted that sour orange is a variable species and a number of different seedling strains of it are recognized. Nurserymen have usually paid little or no attention to this matter, on which attention has especially been focused by the search for stocks resistant to tristeza. Some of the variable results reported in the use of sour orange stocks, for example, for 'Persian' limes and 'Owari' satsumas, may be due to use of different types of sour orange. In the future, especially as scion varieties free of virus become available, it is likely that strains of sour orange will be used according to their adaptations.

3. CLEOPATRA TANGERINE (*Citrus reshni*)—this species came into use in Florida rather late, having been first used by Reasoner Brothers about 1920 as a stock for 'Temple' oranges on highly sandy soil. It attained a small success as a producer of good quality in 'Temples' and

tangelos on such droughty sands and as a substitute for sour orange with kumquats on low ground, but remained a very minor stock until the discovery of tristeza in Florida in 1952. In the search for a stock to replace sour orange for general use, 'Cleopatra' showed up very impressively and is now in third place among Florida stocks. It is nearly as cold-hardy as sour orange, and is very resistant to gumming diseases, scab, and tristeza, and tolerant of xyloporosis and exocortis. It makes good unions with all kinds of citrus scions and produces on them fruit of high quality. It is slower growing in the nursery than other common stocks and trees on it come into bearing a little more slowly, but by the time they are ten years old they have made up the difference in both size and yield, even on deep sandy soils, except possibly 'Valencias'. In many ways 'Cleopatra' seems to combine the good features of both 'Rough' lemon and sour orange. About 12 per cent of the nursery trees sold during the last decade were on this stock, though these were mostly 'Temples' and tangelos. Today it is continuing in popularity.

4. Sweet orange (*Citrus sinensis*)—this stock was extensively used in the first big expansion of the citrus industry after budding became standard in the early 1880's, but gave rise to much trouble from foot rot. This was particularly a problem on the low, fine-textured soils, but was encountered annoyingly even on high pineland. By 1890 sweet orange was no longer in favor as a stock, and when the second big expansion took place on the sand hills of southern central Florida after the big freeze in 1894–95, hardly any stock but 'Rough' lemon was used. When not attacked by foot rot, sweet orange stock produces orange, grapefruit, and tangerine trees larger than on sour orange, though not so large as on 'Rough' lemon, and with fruit quality almost as good as on sour orange. Union of stock and scion is good with all kinds of citrus, but scions are delayed a little in coming into bearing as compared to 'Rough' lemon. The stock is resistant to common scab and tristeza, and tolerant of xyloporosis and exocortis; it lies between 'Cleopatra' and 'Rough' lemon in cold-hardiness.

In the early 1920's, when there was a big postwar expansion of plantings on deep sands, several citrus leaders urged the substitution of sweet orange for 'Rough' lemon stocks on these soils because many old groves on sweet stock were producing good yields of fine fruit. Their arguments made few converts, however. Because it produces a better quality of fruit than does lemon stock, there has been some interest in sweet orange stock for well-drained soils, mostly for tangerines and tangelos, but for these 'Cleopatra' is much more popular. However, sweet orange has never been widely accepted nor is it being propagated in nurseries to any extent today. In theory the seeds of any sweet orange variety may be used with equal satisfaction, but in practice only the seedy varieties produce enough seed to be useful.

Trees on sweet orange stock appear to be much less likely to be affected by young tree decline and blight than those on 'Rough' lemon stock. They will, however, wilt sooner than trees on other stocks during periods of low soil moisture. This stock should be considered in spite of the threat of foot rot. Many groves exist with only minor foot rot problems; furthermore, there is a feeling that after the first few years of the grove's life the problem is lessened to the extent that commercial results can be obtained, especially when good sanitation methods are used.

5. Trifoliate-orange (*Poncirus trifoliata*)—by far the most cold-hardy of the citrus stocks in northern Florida, where the stock becomes dormant early and holds the scion dormant better than other stocks; trifoliate does not confer exceptional cold-endurance in central and southern Florida. Here the warm weather in early winter may delay the dormancy of the stock, and the scion may even be more hurt by cold than on other stocks. It is usually reported as unsuited to high, well-drained sand or to calcareous soil, but well adapted to low, fine-textured soils. It is tolerant of tristeza and xyloporosis viruses and quite resistant to foot rot and other gumming diseases, but is attacked by scab and is the most susceptible of all stocks to exocortis virus (or scaly butt). Fortunately this disease has never appeared in satsumas on trifoliate stock, since this is almost the only stock used for satsumas. It is highly resistant to the citrus nematode although susceptible to the burrowing nematode. So far as is known, trifoliate stock always overgrows the scion markedly, but while some kinds of citrus are dwarfed on it, other kinds make standard trees in spite of what seem to be poor bud unions. Dwarfing occurs, apparently, with exocortis infection. Growth is slow in the nursery, but trees on trifoliate start to bear earlier than on any other stock and produce fruit of excellent quality. Yields are good for the tree size attained. In the past it has been used little in Florida except for satsumas and kumquats although it is used now for other types of citrus in restricted numbers. In California and Australia it has been valuable on certain fine-textured soils for oranges. Because exocortis can easily be avoided by careful bud selection, the resistance to tristeza and the high quality of fruit make it of interest as another possible replacement for sour orange on low soils. It should not be interplanted with other stocks because of the possible transmission of exocortis from symptomless combinations by pruning tools.

About 1880 several nurserymen reported that trifoliate was a better dwarfing stock than the previously used Otaheite-orange, both for oranges and kumquats. In 1892 G. L. Taber began budding satsumas on this stock, with great success, and after 1894 all citrus varieties in his big nursery at Glen St. Mary were worked on trifoliate stock because of its record performance in the freeze.

6. GRAPEFRUIT (*Citrus paradisi*)—rarely or never used today as a stock, grapefruit was considered promising at one time because trees budded on it grew vigorously on low, fine-textured soils and seedling grapefruit trees bore phenomenal crops. Interest declined rapidly because yields on this stock too often were disappointingly low. Some orange and grapefruit groves on this stock are still in production, but the stock has only historic interest today. "Yellow spot," a symptom of molybdenum deficiency, is often prevalent in groves on grapefruit stock on acid soils; although spraying with sodium molybdate can be used to correct this symptom, there is no evidence that yields may, thereby, be improved.

Many other types of citrus are being tried as stocks in the hopes of avoiding virus, mycoplasma, and nematode problems and of finding a solution for young tree decline. In the 1950's, through extensive field search, three trees were found which showed resistance or tolerance to burrowing nematode. These were named: 'Milam', a resistant citrus hybrid of unknown parentage; 'Estes', a tolerant 'Rough' lemon type; and 'Ridge Pineapple', which showed resistance. 'Milam', the only one of the three that has been used with any frequency, is increasing gradually in popularity. (See further discussion in Chapter 7 under spreading decline.)

'Rusk', 'Morton', and 'Troyer' citranges have been used commercially, while today 'Carrizo' citrange is a favorite of many nurserymen and growers in Florida, accounting for upwards of 25 per cent of nursery trees at the present time. Budwood used on the citranges, as with their trifoliate parent, must be free of the exocortis virus. 'Rangpur' lime is gaining more attention as exocortis-free budwood becomes available. 'Palestine' sweet lime (sometimes called sweet lemon), *Citrus limettioides*, is deserving of attention for its yield potentials on sandy soils with xyloporosis- and exocortis-free budwood. The alemow (*C. macrophylla*) is susceptible to tristeza and xyloporosis viruses but this has not been a deterrent to its widespread use in California and Florida as a rootstock for lemons, even though most of the commercial lemon budwood in Florida carries xyloporosis. Unfortunately, wide testing in commercial groves will be needed before complete recommendations can be made concerning these rootstocks and before one or another will take its place as one of the standard stocks of the industry.

In summary, it should be remembered that while the stock does not bear a single fruit, its selection is just as important to the grower as is the selection of the scion variety. No better evidence can be given of the worth of a given combination of stock and scion than the appearance and productivity of a grove of this constitution which has survived the nursery and early growth periods and is in its prime fruiting condition at 15 to 25 years of age.

Budding Procedures

Two important phases of successful budding are choice of a suitable stock and selection of satisfactory budwood. Since the factors influencing stock have been discussed above, the present section will deal only with budwood. Under the Citrus Budwood Registration Program of the Division of Plant Industry, it is now possible to obtain budwood (and stock seed) which is certified free from most of the virus diseases and true to type for the variety. In this program the Division has had the cordial cooperation of state and federal research men, as well as the backing of progressive citrus growers, and this cooperative effort has brought the dawning of a new day in the Florida citrus industry. No person now entering the nursery business on either a home or commercial basis needs to face the uncertainties which were unavoidable to nurserymen of the past or earlier generations, if he will only take advantage of available help.

In selection of budwood, there are two important steps: the selection of the tree from which scions are to be taken and the selection of bud sticks on that tree.

The following points should be considered in selecting the parent tree:

(1) It should be true to the variety, or strain of the variety, in fruit characters, and should be free from any indication of chimeric development, indicated by the appearance in the same tree of two types of foliage, fruit, or growth resulting from a mutation which produces, in the same tree, two or more genetic constitutions.

(2) It should have a record of satisfactory production over a period of at least five years, to even out any difference from year to year due to alternate bearing.

(3) It should be free from systemic (virus) diseases such as psorosis, xyloporosis, tristeza, and exocortis. Absence of visible symptoms is not sufficient evidence of freedom, for many scion varieties may carry infection without showing any evidence of it on one stock and give

abundant evidence when worked on another stock. This is where the above-mentioned Registration Program plays such an important role. It should also be free of propagatable malformations such as witches'-broom.

(4) It should have attained maturity, be in apparently good health and vigor, and have reached a normal size for its age, stock, and soil type.

In selecting budwood from the chosen tree, the following points should be taken into account:

(1) The bud stick should be all of one growth flush, usually the second one back from the tip of the twig, and should have from 5 to 9 good plump buds, not yet starting into growth.

(2) The budstick should consist of strong, well-matured wood, free from scale, and as large in diameter as possible. Well-rounded twigs are much easier for the novice to cut buds from than angular wood, although the experienced budder will have equally satisfactory results with either. Leaves should be cut off at once to conserve moisture in the bud stick.

(3) Budwood should preferably be cut just prior to use and protected from drying out by keeping the sticks in damp sphagnum or peatmoss. However, for budding in early spring, when buds may start into growth on the twigs as soon as the bark will slip, it is advisable to assure a supply of unsprouted buds by cutting budwood before any buds begin growth. The bud sticks should be put at once into polyethylene bags, such as are used for freezing vegetables, and stored in a household refrigerator at between 40° and 45°F. In commercial practice, quantities of budwood may simply be stored in a cool place in damp peatmoss or sawdust wrapped in burlap.

(4) Bud sticks should be labeled at once on cutting them, if more than one variety is used, for it is impossible to identify the variety after it is cut from the tree.

The development of a budded nursery tree, ready for planting in a grove or in the home grounds, may be conveniently considered in five phases:

(1) *The seedbed.* This should be established on well-drained soil free from disease organisms, and should be located conveniently for watering and other maintenance. The site must also be approved by the Division of Plant Industry. After clearing, plowing, and harrowing, the seedbed should be made smooth and level. For large numbers of seedlings, where mechanical cultivation is important, seeds should be planted in rows from 1 to 3 feet apart (depending on the cultivating instrument), spacing the seeds about an inch apart in the row and

planting them an inch deep. When rather small numbers are involved, seeds may be sown fairly thickly in a solid block. No shading of the seedbed is needed in Florida. Trifoliate seed, which is hardy to cold, is usually planted in the fall, but seed of all other citrus is usually planted in spring as soon as danger of frost is past. The soil should be kept moist, but not wet, until the seedlings appear, which should be in about a month; thereafter, water should be supplied as needed to prevent wilting. Maintaining too moist conditions in the seedbed may cause damping-off.

Citrus seeds are unlike seeds of deciduous fruits in that once they have become thoroughly dry they are no longer able to germinate. It is necessary to use seeds which have not been out of the fruits for more than two or three weeks, unless special storage precautions have been observed, and seeds planted within a few days of extraction will be best. Freshly extracted seeds may be stored for a few weeks by washing them, drying them in thin layers in the shade for a day or two, and then mixing them with clean sand in boxes kept in a cool place; or the surface-dry seeds may be stored in a plastic bag in the household refrigerator at around 40°F.

The developing seedlings should be kept free of all competition from weeds, and should receive frequent (bimonthly), light applications of a good garden fertilizer with a 5-5-5 or similar analysis. Spraying is rarely needed in the seedbed. If the seedlings are at least the thickness of a lead pencil at the base by the end of the growing season, they may be lined out in nursery rows. No fertilizer application should be made to the seedbed after September 1.

(2) *The nursery row.* Preferably the nursery should be located on soil of the same type as that in which the trees will be set out, although this is not at all essential. The commercial nursery should be on virgin soil, if possible, or at least on soil which has not previously been used for citrus. If the same site is used for a nursery again, there should be a year or two between such uses in which green manure crops are grown and turned under. Like the seedbed, the nursery soil should be well plowed, harrowed, and leveled, and free from pieces of roots.

Before transplanting the seedlings the seedbed should be well watered to assure the plants being turgid. Not more than a foot of taproot can easily be transplanted, so the seedlings should have their roots cut at a depth of 10–12 inches with a long-bladed spade. Only vigorous plants of about the same size should be transplanted, with all weak plants or those with undesirably crooked roots discarded. Twice as many seeds should be sown in the seedbed as the number of seedlings desired in the nursery rows, to allow rigorous culling in transplanting.

Nursery rows are spaced 3 or 4 feet apart for mechanical cultivation, and the seedlings are spaced 1 foot apart in the row. Care should be taken in transplanting not to let the seedling roots dry out and not to permit taproots to be bent over. Furrows should be deep enough to allow the shortened taproot to be in vertical position. Seedlings should

FIG. 3. A citrus nursery in the lower Ridge area. This view of Ward's Nursery shows also the wind machines used to help protect the nursery from cold damage.

Courtesy Ward's Nursery

always be set at the same depth in the soil as they grew previously, since the root system is adapted to conditions at that depth. Thorough watering immediately after lining-out the seedlings is essential, to settle the fine soil particles in close contact with the fine roots. Fertilizing, watering, and cultivation for weed control should be continued as for the seedbed, so that rapid, vigorous growth may be made. Spraying for control of mites and scale insects, or of scab on susceptible species, will likely be needed. By the end of the next growing season the seedlings should have attained a diameter of at least ½ inch and be ready for budding, although seedlings of 'Cleopatra' and trifoliate-orange may not reach desirable size in the first season and must be left another year in the nursery.

(3) *The budding operation.* A diameter of ½ inch at the base is considered desirable for citrus stocks to be budded, although seedlings of diameters from ¼ to 1½ inches can be worked. Budding can be done only when the bark will "slip," meaning that because of active cambium division the bark can easily be separated from the wood. Budding may be done at any time when the bark slips and unsprouted buds are available, but most budding of citrus in Florida is done in early fall—September and October. Early spring is the next most favorable time for budding. If temperatures are above 50°F., seedlings can be induced to make active cambial growth when the bark sticks tightly, that is, change to a slipping bark condition, by application of nitrogenous fertilizer and water.

Budding in very late summer or early fall is called "dormant" budding, not because the stock is dormant (for budding would then be impossible) but because the scion bud stays dormant all winter if the budding is not done too early. If one waits too late in the fall before budding, the stock will be dormant and the bark will not slip. If budding is done too early, the bud will start into growth in the fall and be very subject to cold injury. Buds inserted in spring will have a fairly long growing season ahead before cold weather and should be well matured by that time. The dormant buds, however, can start growth at least a full month before the spring buds, since the dormant buds united with the stock in the fall while the spring buds need a month or so for union before they can grow.

The type of budding used for citrus in Florida is called "shield budding" because the scion piece is cut in an oval or shield shape. This shield of bark plus a little sliver of wood bears the bud proper in its center, like a boss. A T-shaped incision is made in the stock bark to receive this shield. In most citrus-growing areas of the world, including Florida, the cut is made to form an inverted T so that the shield is pushed up from below. The upright T is used in California and some other areas, however, and there seems to be no inherent superiority of either method. A skilled budder can attain as high a percentage of success with one as with the other, depending on which one he has acquired skill in using. There is some variation in different citrus areas also in the height at which the bud is inserted, but in Florida it has long been considered best to make the stock incision only 2 or 3 inches above the ground. Current research has shown, however, that less foot rot will occur on a susceptible scion variety when the bud is placed at a height of 6 inches above the soil line. The vertical cut should be an inch long.

The shield should be about ¾ inch long, cut from the bud stick so that the inner face is nearly flat and not tapering greatly toward either end from the center. Lifting the corners of the bark at the bottom of the stock incision, gently force the bud up under the bark, taking care that the shield is not doubled back on itself, until it is wholly within the incision. If the bark does not "slip" easily enough for the bud to enter the incision readily, it is often possible to lift the bark and open the incision for its whole length with the budding knife or a probe, and then to insert the bud. The cambium tissue of the stock is now in intimate contact with the cambium tissue around the wood sliver on the inner face of the shield. For successful union, and to prevent drying out of the bud, it is necessary to maintain this intimate contact. Both needs are met by wrapping the stock tightly with a waterproof material. Formerly, waxed cloth strips were the usual wrapping material, but rubber bands and strips of plastic are often used now. Wrapping is started at the base of the bud and carried upward in a spiral so that each successive turn overlaps slightly the one below it. At the top of the incision the wrap is secured by slipping the free end under the last lap. Excessive pressure of the wrap on the bud should be avoided, but there should be no slack in the wrapping material, for the bark must be held firmly against the bud. When budding is done in late spring or summer, it is better to leave the bud itself uncovered, since it may start growth within two weeks and be injured if covered.

Another method of budding which has rather recently gained some popularity in Florida is the use of a so-called "hanging bud." Instead of the usual T-shaped incision, a piece of bark about an inch long and ¼ inch wide is lifted by a single upward cut with a knife held almost parallel to the bark surface. The lower half of this flap is cut off, and the usual shield bud is inserted under the remainder. The bark flap should cover the shield as far down as the bud itself, which should not be covered, and the pressure of the bark on the upper part of the shield holds it in place so that it "hangs" with the lower end free. Wrapping is done in the usual manner for standard shield budding. The advantages of this method are that making the incision and inserting the bud may be done slightly faster, and failure of the bark to slip is of slightly less importance than in the standard method.

About 3 or 4 weeks after budding, (2 weeks if budded during the summer period), the wrap may be removed and the bud examined. If it is still green and callus tissue has formed, the bud has "taken."

If the shield has turned brown and slips easily out of the incision, a new bud may be inserted at another place if the bark will still slip, or rebudding may be done in the spring. If these are the usual dormant buds, the stocks on which the buds have taken may be banked, after unwrapping, with clean soil for the winter about the end of November to prevent injury of the bud by cold. This bank will be removed again in spring after the last danger of injurious cold. Of course, the stocks budded in spring have no need of banking.

(4) *The budlings.* After removal of the banks from dormant-budded stocks, or as soon as union is sure in stocks budded in spring or summer, the stocks should be cut off just above the bud. A sloping cut is made from a point just above the bud downward at about a 45° angle across the stem of the stock. This assures the starting into growth of the bud and conserves for its development all of the water and mineral nutrients taken in by the roots. Amateurs may find it easier to make this cut in two steps, first cutting squarely across the stock an inch above the bud, and then making the finish cut as above. A clean cut, with no ragged edges, will permit more rapid healing of the wound. As soon as the bud pushes out, the stock and scion constitute a "budling," or young budded tree.

A stake should be provided for each budling, so that the developing shoot can be given support and assured of upright habit. This stake, about 4 feet long and an inch square, should be placed close against the stock adjacent to the bud, and be driven a foot or so into the ground. Heavy galvanized wire (No. 8) may be used instead of the stake. When the budling shoot has reached a length of 4 or 5 inches, it should be tied to the stake with soft, heavy twine, fastened loosely enough so that the expanding stem tissue will not be constricted. As the budling continues to grow, additional ties should be made at about 10 and 15 inches above the union. By the end of the growing season the scion will no longer need support and the stake can be removed before digging.

Care of the budling should be the same as has been discussed for the nursery already, with watering, fertilizing, cultivating, and spraying carried on as needed for healthy, vigorous growth. The budlings should be examined quite frequently for stock sprouts so that these may be removed before they have a chance to compete seriously. Quite young, tender sprouts can simply be rubbed off; older ones should be cleanly pruned off.

(5) *Digging the budling.* At the end of one growing season after budding, the budling should be ready for transplanting to the grove, with a diameter just above the union of at least ⅝ inch. If this caliper

has not been attained, the tree may be left for another season. The first step in digging is to "top" the tree; that is, cut off the top, at a height of about 2 feet. If the tree has already formed branches from about a 20-inch height upward, these may be cut back to 6- or 8-inch stubs. Then the lateral roots are cut at about a foot from the trunk all around the tree with a shovel, except that the cut must be made midway between adjacent trees in the row. Finally the roots are undercut below the tree at a depth of 12 to 18 inches with a long-bladed nursery spade. Using this as a lever with one hand, one pulls the tree from the ground with the other.

The greatest hazard in transplanting citrus is drying out of the roots. The excess soil clinging to the roots in digging is shaken loose and the roots are at once covered with wet sawdust, peat moss, or other water-holding material. The top of the tree, devoid of, or retaining a few leaves, is then covered with a tarpaulin or similar material to cut off sun and wind and maintain a humid atmosphere. The budling can now be taken to the planting site. Avoid injury from cold winds during the winter season by use of covered conveyances.

Topworking

The term "topworking" means the operation of changing the scion variety of a tree which has already formed its scaffold branches, by budding or grafting these with another variety. Either a seedling or a budded tree may be topworked, although there is rarely occasion to topwork seedling citrus trees. The process is mostly employed to change an undesirable variety or strain to a desirable one, without losing the advantage of its established root system. The topworked tree can be in production in two years and will be much larger than would be the case if the old tree were pulled out and a new nursery tree planted. A variety may be undesirable because it was unwisely selected in the first place or because changing market conditions make it less profitable than was anticipated. One may discover when his trees commence to bear that he has a poor strain of the desired variety, perhaps simply unproductive or perhaps bearing fruit of poor quality. In either case the tree can be topworked without sacrificing more than two or three crops. In the home planting the owner may wish to have several varieties of fruit borne by a single tree, to conserve space. It is only feasible to grow a single variety in the nursery row, but once well established in the grove, more varieties can easily be added by topworking. Thus, early, midseason, and late oranges may be produced on the same tree, or grapefruit and tangerines may be borne also on an orange tree.

Topworking can be done in several ways, using budding or grafting as the situation warrants. The easiest way is to cut back the tree to short stubs of its scaffold branches and to bud the shoots which develop from these branches as if they were seedling stocks in the nursery. A second possibility is to insert buds into bark incisions on the branches just below the points where they will later be cut off after the buds have "taken" satisfactorily. With older trees it is usually necessary to thin the bark by paring or scraping it before budding can be done, and a larger bud shield is preferred than for nursery budding.

Either the usual T-incision (inverted or upright) can be made, or a single oblique cut made from one side, the side- or hanging-bud technique. In the latter case the flap of bark must have a notch cut to avoid covering the bud proper, or "eye." Wrapping of buds is carried out as for nursery budding.

Another method is to cut off the scaffold branches as above and then insert scions at once, instead of waiting for sprouts to reach budding size. The branch stub may be split clear across and a cleft-graft made, or if the bark will slip, the scion may be inserted under the bark in a bark-graft. If the grafting is successful, a season of growth may be saved over budding sprouts; if it fails, the sprouts can still be budded.

In cleft-grafting it is of utmost importance that the cambium layers (the point where bark meets wood) of stock and scion should match exactly. The bark of a 3-inch stub is sure to be much thicker than that of a ¼-inch scion, so that the latter must be set slightly inward from the edge of the stock. Of course, only the cambium on one side of the scion can be matched. Scions should be chosen of well-matured twigs, usually not the last flush of growth, with plump but dormant buds, and should be 5 or 6 inches long and from ¼ to ½ inch thick.

After the stock is split to a depth of 3 or 4 inches, the cleft is held open by a wedge (either the one provided for the purpose on a grafting iron or a special one) in the center of the stem to the necessary width to allow easy insertion of the scion. The sides of the cleft should be carefully trimmed to a smooth, even surface. Then the basal 2 inches of the scion are cut to a tapered wedge, very slightly thicker on one side than the other. This wedge is inserted in the cleft at one side of the stock, with the thicker edge next to the periphery, so that cambium layers match. If the stock is less than 1 inch in diameter at the cut surface, only a single scion is likely to be used. With larger stocks a second scion is inserted on the other side of the stock similarly. Then the center wedge holding the cleft open is carefully removed, allowing the pressure of the stock to hold the scions firmly in place. In the case of large stubs, either sturdy scions must be used or this wedge only partly removed so that it prevents crushing of the scions. After completing the graft, the stub is wrapped with tape and the cut ends of stock and scion are covered with grafting wax or asphalt emulsion to prevent drying out of cambial tissues and entrance of disease organisms.

Two other possible methods are both rather unsatisfactory. They involve cutting the tree to the ground and either crown-grafting the stump or budding sprouts after they have developed. In both cases the scaffold structure of the old tree is lost, the healing of the one big stump is slower and less certain than that of several branch stumps, and the chances of developing a satisfactory new tree are reduced.

Trees seriously hurt by cold may be killed well below the forks, perhaps even below the bud. Such trees must often be cut off close to the ground. Either cleft-grafting, bark-grafting (usually called crown-grafting in this case), or budding of sprouts may be employed to make a new top, but serious consideration should be given to removing and replacing trees so seriously hurt. While several scions may be inserted, it is best to select the strongest one after they have made some growth and use it to form the new top as if it were a nursery budling, pinching back and eventually eliminating the other scions. If all are allowed to grow, they usually form very acute crotches which permit the whole limb from one or more scions to split off under a heavy load of fruit. A tree should be cut below the main forks only by necessity.

The new growth developing naturally from latent or adventitious buds on branch stumps, or from buds inserted under the bark of these branches, is exceedingly lush and soft because of the very large amount of food and water supplied by the old stock. This growth is not very firmly attached to the old branches for two or three years and is easily broken off at its point of origin by wind. To prevent this the bud growth should be supported by being tied to stout stakes driven into the ground beside them or nailed to the old branch stubs. If budding is done before cutting back the scaffold limbs and new shoot growth has pushed out from the buds too, special care is needed in cutting off the limbs. A preliminary cut can be made several inches beyond the new shoot, and then a second cut made close to it more easily.

No tree should be topworked unless it is vigorous and healthy. If the trunk and limbs show evidence of disease or serious injury, they cannot be expected to produce a vigorous new top or to give it a long life. Grapefruit trees are topworked very successfully because of the sturdy and well-formed scaffold branches which they make. Usually all varieties of sweet orange can readily be topworked, also. Tangerine trees, however, have not proven very satisfactory for topworking, since the new top often breaks off at the point of union with the old trunk when the first heavy crop is borne. As a rule it is better to remove tangerine trees and replace them with nursery trees than to topwork them.

When a tree has been topworked by budding to another variety of the same species, it is often very difficult to distinguish the shoots of inserted buds from shoots developing from the old branches or trunk. For this reason it is important to inspect the topworked trees at frequent enough intervals to remove stump sprouts while they are still very small. Leaves of orange can easily be distinguished from leaves of grapefruit, but there will be no way of telling one variety of orange or grapefruit from another; and even leaves of the various mandarins may not easily be distinguished from those of oranges in lush sprouts. The careless grower may end up pruning off the buds he inserted if he allows the stump sprouts to grow large too.

One serious problem in topworking is the possibility of injury to the bark and cambium of the remaining trunk and scaffold branches from sunshine when the tops have been cut off. Growers have learned by bitter experience that sunscald of the limbs of defoliated trees will result in partial or complete death to the system. Bark previously shaded by the canopy of foliage is not adapted to receiving the direct rays of the sun and must be shielded from such insolation when cutting back is done at any time other than just before the spring flush of growth. This is often done by whitewashing the trunk and limbs with a hydrated lime-water mixture with a "sticker" added at once after removal of the top of the tree, to reflect sunshine and thus prevent excessive heating of delicate tissues; or it may be accomplished by wrapping the exposed parts with burlap, Spanish-moss, or kraft paper. Cement paints, such as are used to waterproof the outside of cement-block buildings, make excellent whitewashes that will generally remain effective a whole year from a single application. It is a good practice to topwork only one-third of a top each year, so that the foliage of the other two-thirds serves to continue the natural protection of the bark from the sun. In this way the whole top is renewed in three years, there is never a period of no fruit production, and there should be no sunscalded trees to pull out. Great care must be taken in the second and third year of this program to avoid breaking off the young growth from the previous year's topworking when removing the rest of the top. However, if due care is taken (whitewash may need renewal before the new foliage shades the trunk) it is perfectly possible to do all the topworking at one time. In either case, a good job of topworking on a healthy trunk should produce a top with fruit production about equal to the original top in about five years from starting the work.

Inarching

Sometimes the root system of a tree is seriously injured or impaired by fungi or burrowing animals, so that loss of the tree is threatened. Usually it is best to remove such a tree and replace it with a healthy one, perhaps budded on a stock more suited to the soil conditions. However, in the case of trees with historic value it is often possible to replace the diseased or injured roots by inarching a vigorous seedling into the trunk. One or more seedlings of size suitable for budding or a little larger are planted as close as possible to the base of the ailing tree. As soon as they have become well established and have a stem at least 2 feet high, they may be grafted into the trunk of the tree. One common technique is to make an incision in the trunk as if for shield-budding and to insert the obliquely sharpened end of the seedling stem into this. Another is to cut a slot in the bark of the trunk of the same width as the seedling stem, and after cutting away a thin slice of bark and wood from the latter, press it tightly in place so that exposed cambium of seedling and trunk are in contact. In either case the cuts must be made with care to assure matching of the two components to be joined, and the seedling stem must be held firmly in position by nailing a slender brad through it into the tree trunk until union takes place. Presently the new seedling roots will take over the functions formerly performed by the original root system, and the rejuvenated tree may have a long increase of life. This method is not at all recommended for commercial use, however, as it is too slow and expensive to be profitable.

4. Climate and Soil in Citrus Growing

Climatic Considerations

CLIMATE refers to the general or average conditions of a given area as regards the various atmospheric phenomena, while *weather* refers to the transient conditions of atmospheric environment for this area. In other words, weather is the climate of the moment, and climate is the average weather. Both the weather at any given time and the climate over many years are of importance to the citrus grower. Their influence on citrus growing may be either obvious or obscure; they may make the growing of some or all kinds of citrus fruits entirely impossible in some areas, while in others they exercise an almost imperceptible restraint on commercial production. No major citrus-growing region of the world is without disadvantageous climatic factors of one sort or another. In spite of the fact that various citrus species flourish in different parts of this state as though indigenous, the Florida grower is forced to recognize the limits set by climate; indeed, he will probably spend more waking hours—often when he should be sleeping—in worries concerning climate or weather than in the administration of his production program.

Five phases of climate (or weather) are of interest because of their influence on tree growth and crop production: temperature, rainfall, relative humidity, wind, and sunshine.

Temperature

The temperatures found in the citrus-growing region of Florida are in general such as would be expected in a subtropical climate. During the period from April to October they are moderately high, but these summer temperatures, though long continued, do not reach as high values as are experienced in many more northerly areas. The highest summer and lowest winter temperatures occur in mid-continental regions, whereas the presence of either the Atlantic Ocean or the Gulf of Mexico within 75 miles of any point in Florida serves to moderate both summer maxima and winter minima. The highest daily temperatures in summer are usually from 93 to 95°F., with higher temperatures at irregular intervals which rarely ever reach 100°, although an all-time maximum of 107° was once recorded in the state.

From October through March lower temperatures prevail. Minima below 32° are expected every winter and invariably occur in the northern part of the state, while they are not reached every winter in the southern part. Only the Florida Keys are completely free of frost, but the minimum temperatures are lower as one goes from south to north in the state. It usually follows also that the lower the minima in a given area, the longer will be the duration of low temperatures and the more often they will occur. Thus the northern part of Florida not only has lower minima in any winter than southern areas, but the cold spells last longer and come more frequently.

Injurious low temperatures may occur under two sets of conditions —those of a *frost* or those of a *freeze*. In both cases the temperature of the air or of plant tissues goes below 32°F., but for different reasons; and the methods of preventing plant injury will be somewhat different under the two conditions.

Frosts are local occurrences, the cooling of the air and plants resulting from radiation of heat from soil and plants out into space. As the soil cools by radiation, the air above it gives up its heat to the soil and so is slowly cooled upward, with lowest air temperatures next to the ground. Rapid loss of heat by radiation occurs only on clear, cloudless nights, for clouds reflect back this heat in large measure. On a calm, still night the air forms a series of layers progressively warmer from the ground up—spoken of as a stratified condition—but if there is a light breeze these strata become mixed. Then the warmer upper air mixes with the colder air next to the ground and prevents it from reaching as low temperatures. Thus frost damage is most likely to occur on still, cloudless nights, and frosts may occur anywhere on the mainland of the state. Muck soils radiate heat faster than sandy soils, and for the same latitude danger of frost is greater on muck. Elevated areas are less subject to frost than adjacent low areas because of air drainage.

Freezes are always general, not local, because they result from the importation to the state of large masses of air at subfreezing temperatures. These masses of frigid air often follow certain channels for topographic reasons, somewhat as water flows in valleys. Because the air is at the same temperature from top to bottom of the moving mass, there tend to be no differences between temperatures on high and low ground, at least on the first night of a freeze. Whereas a frost usually follows a warm, sunny afternoon, when soil and tree have stored considerable heat, the afternoon before a freeze is usually cold and windy and may even be cloudy. A frost lasts only for a night, although it may recur the next night, but a freeze usually has at least a three-day period.

The first night is always cold and windy, but rarely causes serious damage, although there may be a period of calm a little before sunrise which allows the air to stratify, with possible danger then. During the second day there is usually little warming of the air or trees by the sun and cold air continues to move south and chill things more. During the second night the wind usually falls soon after sunset, and the stratifying air may reach a dangerous low temperature very soon in low areas, while adjacent high ground maintains higher temperatures. On the third day there is usually a shift in the wind to start replacing the cold air with warmer air from over the ocean, and temperatures begin to rise from the south northward. Under the usual conditions of freezes in Florida, therefore, the second and/or third nights, after ground and trees have become cold and the wind has ceased, are most disastrous. This is further accentuated if daytime maxima do not exceed 60°F.

Freezes may occur in Florida anytime after November 15 until March 15. The most severe injury results when an early winter freeze is followed by a period of warm weather sufficient to initiate new growth, and this in turn is followed by a second freeze the same winter. Such was the pattern of the freeze of 1894–95, still spoken of as the "big freeze." Trees were defoliated and fruit was frozen, but wood damage was not severe from a freeze in early December of 1894. During January the weather was mild, trees put out vigorous new shoots, and growers felt that they had come through the experience in good shape. In this condition of tender growth, however, the trees were killed to the ground by a second freeze in early February of 1895. In January, 1940, a freeze of several days duration caused much loss of fruit and considerable injury to the branches; yet, because there was no later visitation of severe cold that winter, the new growth in February following the freeze developed normally and the trees were practically back to normal condition that summer. The freeze pattern of the winter of 1957–58 was again one of repeated cold waves interspersed with periods of sufficient warmth for renewal of growth, and again damage was severe in many areas. The "Big Freeze of December 1962" was the most damaging of this century. Cold temperatures occurred from December 11 to 16 with hard freezes on the 13th and 14th. Other freezes have produced lower minimum temperatures in low ground locations; the convection (blowing) freeze of 1962, however, resulted in low temperatures and severe damages even on high ground locations. Less damage occurred in the Indian River area, and in southeastern Polk County and Highlands County. Various sections of the state have suffered losses from freezes in years in which a general freeze did not occur.

Wishful thinking often allows growers to be lulled into the comfortable feeling that because groves are now in generally better health than before ways were known for correcting debilitating deficiencies of nutrient elements, the trees should be able to endure freezes now without injury. Florida's history of broken winters and their effect on citrus trees is an open book for those who can and will read it.

Cold temperatures limit the northward expansion of the citrus belt and are the most adverse climatic factor with which the Florida citrus grower must contend, but high temperatures cause some trouble too. As just seen, relatively high temperatures (in the 70's) during December and January may encourage growth and thus make trees more easily injured by later cold. In March and April, high temperatures, coupled with lack of soil moisture and hot, dry winds, increase transpiration until trees often go into permanent wilting. When such conditions are prolonged into May, although not serious enough for such wilting, an excessively heavy "June drop" of fruit may be expected. Normal high summer temperatures in June and July, with high humidity and intermittent cloudiness, may accentuate sunscald and spray burns. Warm weather during October and November, particularly if nights are warm and rainfall is much above normal, usually results in delayed development of internal quality and external color, and in increased tendency for dropping of fruit, as well as in delaying maturity of tree tissues and thus increasing susceptibility to injury from winter cold.

Temperatures vary slightly according to latitude and to proximity to the Atlantic Ocean or the Gulf of Mexico. Minimum temperatures are lower as one goes from southern to northern Florida, and as one goes from the coast inland if elevation does not change. Locally temperatures are influenced by such variable factors as: elevation, soil type, air channels, bodies of water, surrounding vegetation, extent of grove areas, presence of windbreaks, presence of hammock trees within a grove, and cultivation practices. Even though no way is known to prevent the occurrences of periodic freezes in Florida, the grower can minimize their effects by proper selection of site and stocks and by maintaining satisfactory cultural programs.

The temperature range for growth of most of the commonly cultivated kinds of citrus fruits is from 55° to 100°F. Within this range the tree is able to develop and fruit, with best growth occurring around 85°, or from 80° to 90°F. The processes of fruit maturation, including production of sugars and development of rind color, reach their highest perfection in the lower portions of the growth range. Winter temperatures ranging from 35° to 50°F. approach the ideal for dormant tree condition. Temperature ranges of 100° to 130°F. and 32° to 55°F. are considered, respectively, high and low endurance ranges. Very low metabolic activity occurs within the plant, no active growth is present, and no damages can be observed.

Injury from low atmospheric temperatures usually begins to be evident when the air temperature gets below 32°F., although under frost conditions the temperature of fruit or leaf may reach the point of injury before the air temperature is down this low, and cold, dry winds may injure tissues while the air is still above 32°. Only very tender foliage or flowers are injured by such temperatures. Fruit is damaged when its tissues reach the freezing point, which may be anywhere from 30° down to 26°, depending on variety and maturity. How long it takes a given fruit to reach this temperature after the air temperature does depends on size, location on the tree, rind thickness, and initial warmth of the fruit. In commercial practice it is considered likely that fruit may show injury after exposure for two hours to air temperature of 28°, but individual fruits will vary widely in this respect, depending on the variable factors enumerated above.

Fully grown leaves may begin to show injury (evident as dead areas and/or as defoliation) after being exposed to from 2 to 4 hours of air temperature of 27° or lower, and because of the small volume of a leaf compared to its area, there will be little variation because of stored heat; however, location on the tree exercises an influence on foliage injury. Mature, dormant orange trees have been reported to survive 10 hours below 25° and 1½ hours at 16° with only 10 per cent of the leaves lost from cold. Previous temperature conditions play an important role in the endurance of cold by trees.

Dormant twigs and limbs may be injured at temperatures ranging from 27° down to 16°, depending on size and maturity of wood and length of exposure to cold, as well as on previous temperatures. Even the most severe freeze known in the citrus-growing areas of Florida will not usually kill the older trees completely, if it occurs only once in a winter or if no warm weather intervenes between successive periods of cold. Because their tissues are less mature, their bark is thinner, and their heat capacity is lower, young trees may be killed outright by a single severe cold spell. Older trees, even though not killed entirely, may have sufficient killing of limbs that they are no longer economically useful, because new trees could be brought into production as soon as, and at less expense than, these frozen trees could be rejuvenated. Often serious damage to trees from cold temperatures is evidenced by "frost or cold cankers" on scaffold branches and trunk, particularly in the crotch area. Cambium, active at the time of freeze, is killed while the surrounding tissue is not harmed; the tree may be left in such a weakened condition that it is not capable of producing satisfactory crops of fruit. Large cankers, on the upper side of scaffold branches, may cause the branches to break under the weight of fruit. Even though the cankers do not become progressively larger, there is no way to eliminate them except by removing the affected branches.

The amount of injury caused by excessively low temperatures will vary, therefore, with the following factors: species and variety of scion and of stock, the condition of the tree with respect to vigor and degree of dormancy, the minimum air temperature reached, its duration, the exposure of the tree to radiation, the conditions of soil cover, and the interrelations of temperature with other climatic factors prior to, during, and following the low temperature. Other things being equal, the vigor or health of the tree, resulting from previous cultural practices of the grower, plays a dominant role in minimizing injury from cold. Trees in good health, with no mineral deficiencies and free from serious insect infestations or disease infections, endure cold much better than weakened trees.

If trees are in well-matured or dormant condition, the trifoliate-orange can endure the lowest temperatures of any of the commercial citrus kinds, followed in descending order by kumquat, calamondin, sour orange, mandarin, sweet orange, grapefruit, shaddock, lemon, lime, and citron. The range is thus from the trifoliate-orange of the temperate zone to the citron of the tropics. When trees are in active growth flush, however, there is little if any difference in ease of injury by cold between the various kinds of citrus fruits. The big factor of difference is usually the readiness with which different citrus species tend to go into or come out of the dormant condition during the winter months.

The trifoliate-orange, so hardy in most winters, may be seriously injured by early freezes of slight severity when the temperatures in the early fall have been unusually high. Apparently, lower temperatures are needed by trifoliate-orange to induce dormancy than are needed by other kinds of citrus fruits, which may be much less injured by the same freeze. On the other hand, trifoliate-orange requires higher temperatures to initiate growth than most other citrus kinds, and so is later than they in starting its buds in spring, thus escaping damage from spring frosts and late winter freezes which may injure the others.

The kumquats also are late in pushing out buds in spring, and so also tend to escape many cold injuries suffered in late winter and early spring by grapefruit, sweet orange, mandarins, and even sour orange, due to their tendency to start bud growth following a few days of warm weather. Within this group the 'Cleopatra' withstands cold better than other tangerines, but the satsumas are the outstanding example of cold-hardiness among the mandarins. Lemons, limes, and citrons respond very readily to mild weather, and usually have flowers, new foliage, and young fruits at almost any time in the winter. This lack of dormancy renders them increasingly subject to winter injury in the order named.

Even when these species are used as stocks, cold-susceptibility must be considered, not so much because of possible killing of the stock itself as because of its influence on the degree of dormancy of the scion. There is no exact correlation of cold-hardiness of a species with the cold-hardiness of a given scion worked on it as a stock, and there may be much variation in the effect of different varieties of the same species as stocks. However, 'Rough' lemon, the most common stock of the Ridge area, is much less satisfactory on well-drained sandy soils of the northern citrus areas than sour orange. Still farther north the trifoliate-orange gives the greatest cold-hardiness, although attention has been called to its less satisfactory ability to make a hardy tree in areas where winters have alternating periods of cold and warmth.

Seedling sweet orange trees are considered to be able to endure more cold after they reach good bearing age than are budded trees of the same age. The same is probably true of grapefruit trees. It has also been noted that varieties within a species tend to be more cold-hardy if the fruit is seedless and decrease somewhat in hardiness with increase in seediness. Probably this is a nutritional phenomenon, reflecting a translocation of mineral nutrients to the developing seeds, leaving the twigs somewhat deficient for a time.

The importance of the duration of subfreezing temperatures is often overlooked by growers, who simply observe the minimum reached. A recording thermometer will indicate better the true temperature situation, for a minimum which causes no injury when it persists for only half an hour may do serious damage when maintained for four hours.

Two environmental factors which influence cold-endurance by their effect on the temperatures actually reached by foliage or fruit are the presence or absence of overhead and ground cover. Trees which are somewhat shaded by pines, palms, or other thin-foliaged trees are always less injured by low temperatures than trees fully exposed to the sky. Indeed, because of this observation, during the freeze of 1886 slat shade was constructed over some orange groves. The shading trees reflect back much of the heat radiated by the citrus trees, which would otherwise be lost to space, and so prevent them from reaching as low temperatures as trees fully exposed to the sky. On the other hand, injury is usually greater when citrus trees have the ground under them covered by grass, weeds, or mulch than when the soil is cultivated. In this case cultivated soil radiates heat more freely to the trees than soil covered with vegetation and so prevents their reaching an injurious temperature as soon.

Windbreaks may be helpful or detrimental in connection with cold periods. A windbreak on the northwest side of a grove slows down the movement of cold air into the grove in a freeze. The stream of cold air flows up over the top of the windbreak and only slowly descends to ground level again, leaving a relatively quiet area in the lee of the windbreak. Heating the grove is more effective in this lee space, since the warmed air is not carried away so rapidly as in a grove with no windbreak on the windward side. On the other hand, a windbreak down-slope from the grove may prevent air drainage on a frosty night and thus cause more cold injury than if it were not there. If the windbreak is both on the northwest side and lower on the slope, occasional openings in it will permit slow drainage of cold air while still giving considerable protection in freezes.

The kind of weather during the weeks preceding a freeze has an important effect on the amount of injury done, through its effect on tissue maturity. If temperatures during the preceding month have been down in the 40's continuously, especially if the weather has been dry, tree tissues will be well matured and will endure quite severe cold with little injury. The absolute minimum endured will vary, of course, with the particular scion and stock involved. However, a prolonged drought may put trees in a condition of water deficiency which renders them more subject to cold injury than trees with adequate moisture content. Attention has already been called to the undesirable effect on cold-hardiness of a week or two of warm weather (60–70°F.) just before a freeze, or even a frost.

Rainfall

The amount of moisture made available by rainfall is an important consideration in the production of any horticultural crop, and citrus fruits are no exception. The total amount of rainfall annually is not the only item of importance in determining adequacy; in addition must be considered the distribution of the rainfall during the year, its seasonal fluctuations, and its intensity.

The average annual rainfall within the Florida citrus belt is approximately 52 inches. The larger portion of this precipitation falls during the rainy summer season, from May to September. During this period the natural rainfall usually takes care of the needs of citrus trees for moisture and often provides an excess over what is needed. From October through April, and occasionally through May or early June, the rainfall is often insufficient for the needs of the trees.

The most critical period is usually in the spring, particularly during the months of March, April, and May, and especially if rainfall was deficient the preceding fall. The two periods in the annual growth cycle of a citrus tree when it is most sensitive to soil moisture deficiency are in early spring when the new flush of growth is tender and fruit is setting, and in late spring and early summer when fruit is rapidly increasing in size. Deficiency of available soil moisture during the period of fruit setting may cause abnormally heavy shedding of young fruit. Tender new shoots easily wilt if soil moisture is in short supply, and seem to have a prior claim on moisture within the tree over young fruit. Deficiency of soil moisture in May and June may prevent citrus fruits from reaching good size later. Shortage of rainfall during October and November is not likely to prove critical unless the fruit is wilted at the time of picking or the trees experience an excessive wilting of foliage. Low temperatures during the winter months decrease transpiration rates of leaves, so that a given amount of rainfall goes further toward meeting tree needs than the same amount in summer.

While the average annual rainfall is about 52 inches, the amount which falls in any individual year may vary from 37 to 84 inches. In addition the proportion of the annual precipitation which falls in any given month also varies from year to year. This annual and monthly variation gives a dynamic quality to the rainfall; it would take about 1,000 years to experience all the possible variations in rainfall pattern which might occur in Florida. Obviously it would be impossible for any one grower to operate for a sufficient time to profit by his short period of rainfall experiences. The static climatic picture gives the impression that predicting rainfall adequacy should be easy, but the dynamic weather pattern makes it very difficult to do so.

The intensity of rainfall, that is, the amount falling in any 24-hour period, is also of importance. This amount may vary from a "trace" (less than 0.01 inch) to as much as 16 inches. A monthly rainfall of 6 inches may all have come in one heavy shower, or may represent a dozen rainy days with half an inch each. It makes a big difference to the trees whether one or the other was the case. Rainfalls of one-tenth inch or less are of little use to citrus trees, except as they occur along with humid, cloudy days which reduce transpirational losses. The precipitated moisture evaporates from the soil surface without ever affecting the soil moisture. A rainfall of one-half inch may wet a sandy soil to a depth of 6 inches, but its effect is temporary because soil moisture to this depth is readily lost by evaporation. Probably only 10 per cent of such a rainfall is actually used by the trees. Rainfalls of from 1 to 3 inches are ideal for the Florida grower, since they wet the soil deep enough to supply moisture over a long period, yet are unlikely to provide excess water which percolates beyond root depth. Of course several consecutive days of half-inch rainfall, with no sunny days for evaporation between them, may also provide satisfactory soil moisture conditions. The very heavy rainfalls accompanying hurricanes, which may run as high as 16 inches in one 24-hour period, are very harmful because they supply far more water than our sandy soils can hold, and consequently they cause serious leaching of soluble nutrients. Any water which percolates to a depth of much below 5 feet in well-drained sandy soils is usually unavailable to the trees.

Relative Humidity

The ratio of the amount of water vapor in the atmosphere at any given time to the amount which the atmosphere would contain at the same temperature if saturated is termed the relative humidity and is always expressed as a percentage. It in no way indicates the amount of water actually present in the air, but it does exercise an important influence on tree growth and health, as well as on the comfort of the grower.

The relative humidity of Florida varies normally from nearly 100 per cent at night and early in the morning to an average low of 40 per cent in midafternoon on clear, sunny days. Of course it is also 100 per cent whenever it is raining. There is little seasonal variation, so that daily average relative humidity of 72 per cent is fairly constant throughout the year.

This high relative humidity of Florida is advantageous for tree growth, for it decreases the transpiration rate for any given temperature and so makes for greater economy of water use by citrus trees in comparison with regions of low relative humidity. However, it has some disadvantages, too, for it is chiefly responsible for the many fungus diseases which cause so much damage to fruit and trees in Florida. Citrus scab, melanose, stem-end rots, and other diseases are much more serious under humid than under arid conditions.

High relative humidity, especially in combination with abnormally high temperature, is the cause of much of the poor textural quality, excessive "puffiness," and lack of bright color sometimes noted as citrus fruits mature in Florida. The difficulty in curing lemons in an area of high humidity was an important factor in the decline of the early lemon industry here, and would still be a handicap to a revived production of lemons for fresh market.

Wind

Three types of winds occur with some regularity in Florida and are rather detrimental to production of citrus fruits. Southeasterly winds, which characterize the spring months (especially March and April), increase the rate of water loss from trees by accelerating the transpiration rate and decrease the available soil moisture by increasing the rate of surface evaporation. Soil moisture is usually not abundant at this time and the new shoots transpire faster than they will when they mature, so that the added stress due to these winds is particularly a problem at this time. If rainfall has been unusually low during the winter and spring, severe injury to tree or crop may result.

The hurricane season extends from June 15 to October 15 and is a period when very strong winds may be felt in the citrus area. These winds of hurricane velocity may vary from 60 to 120 miles per hour, depending on how close the center of the storm passes by a given area, and are capable of causing severe physical damage to buildings and equipment as well as to the grove by uprooting trees and shaking much fruit to the ground. Even if the trees and fruit remain in place, the fruit may be badly scarred by rubbing against branches, so that its grade is seriously lowered. This may cause more loss to the citrus industry, on the average, than direct loss of fruit and trees, for it is rarely that any one grove experiences the full intensity of hurricane winds, but a hurricane crossing the state anywhere will be felt as strong, fruit-bruising winds everywhere in the citrus belt.

Freezes always are brought to the state by cold, northwesterly winds, occurring any time between November 15 and March 15.

Sunshine

Sunshine contains several components, notably heat, light, and ultraviolet rays. Light, all-important to green plants for photosynthesis, is never deficient for good plant growth in Florida. Heat has already been discussed under Temperature. Ultraviolet rays are absorbed by the atmosphere to a much greater degree under the humid conditions of Florida than in more arid regions, resulting in less bright color development of oranges and mandarins. For this reason artificial coloring of these fruits is practiced more in Florida than in other states.

Soil Considerations

Soils provide anchorage for plants and furnish the water and mineral nutrients required for plant growth. In the way they serve these functions the native soils of Florida show wide variations. These soils have been grouped by scientists into "series" on the basis of origin, color, structure, and other characters, and these are further subdivided into "classes" according to the texture (particle size) of the surface layer. *Astatula* (formerly classified as *Lakeland*) *fine sand* identifies the series (*Astatula*) and class (fine sand) of the most important soil used for citrus production in Florida in the well-drained areas of the state. Such grouping makes possible the classification of soils as naturally occurring bodies and the study of soil characters with regard to their usefulness to horticultural industries.

The horticulturist has long recognized the extreme importance of certain soil characters and has selected planting sites with due regard for them, even when he knew nothing of the name by which the soil was classified—indeed, long before there was any such classification. Drainage of water is perhaps the most important of these characters, and in itself can be used for grouping soils. But in the early years of the Florida citrus industry great reliance was placed on the type of vegetation growing naturally on different soils.

Soils Classed by Native Cover

The early citrus grower noted that the native flora not only served as a rough index of soil drainage but also gave indications of the natural fertility of the soil and the relative degree of cold to be expected at that location. In the well-drained areas the following types of land were distinguished, being listed in descending order of desirability for growth of citrus trees:

1. *High hammock land* maintained a heavy hammock growth of hardwoods such as live oak, magnolia, hickory, and dogwood. While "high" in comparison with low hammock, this type of soil was actually not elevated very greatly.

2. *High pine land* sustained good stands of longleaf pine, often with some scattered red oak and post oak and little underbrush.

3. *Blackjack oak land* supported turkey oak (sometimes called blackjack) and scattered pines. It represented a poor grade of high pine land.

4. *Scrub,* comprising areas of coarse sand with low organic matter, reflected its low soil fertility in its poor stand of sand pine, turkey oak, and other shrubby oaks.

High hammock and high pine lands proved very well suited to citrus culture, while blackjack oak land has never been found very satisfactory and is often treacherous because of cold pockets. The scrub will probably always remain in its native condition for it has little possible use in agriculture. Scattered through the high pine lands are limited areas of coarse white sand of the same character as scrub land. These small areas, called "sandsoaks," are quite as unsatisfactory for citrus production as large scrub areas.

On poorly drained soils the following types of land were recognized, again given in order of decreasing suitability for citrus planting:

1. *Low hammock land* bore a heavy growth of hardwoods, especially live oak, and of cabbage palmetto. Difficult to clear and to drain properly, it has produced some of the exceptionally good coastal groves of the state. The old, so-called "hammock groves" were established with the underbrush cleared out but many of the larger hardwoods and palmettoes left in place, and were notable producers of fine crops of fruit. A few of these groves can still be seen along the upper East Coast.

2. *Flatwoods land* is low and level, with scattered longleaf pine predominant in the northern part of the state and slash pine in the southern part. Frequently saw-palmetto and wiregrass cover the ground and a hardpan underlies the surface, making drainage very poor. In general these soils are very poorly adapted to growth of citrus trees.

3. *Bayheads,* originally supporting stands of bald-cypress and other trees which endure inundation, are areas of standing water, often associated with well-drained soils at the lower elevations. They are entirely unadapted to citrus culture.

The present-day horticulturist is unfortunately no longer able to make use of natural vegetation as an index of suitability of soils for citrus planting. Most areas which originally supported tree growth have been cut over, especially on soil types reasonably well adapted for citrus culture. There are areas of cut-over blackjack oak land, of flatwoods, and of scrub which have not changed greatly from their native condition, but these are the least desirable soils for planting citrus trees.

Soils Classed by Drainage Character

With the native flora no longer available as a guide, the grower must base his choice of land primarily on two characters: water drainage of the soil, with its effect on depth of rooting, and air drainage of the land for natural cold protection. Today the soils of many areas of the state have been well mapped, so that it is easily possible to ascertain the exact soil type or types in prospective locations within these areas. Knowledge of these soil types and their characteristics is important to the modern citrus grower. Based on drainage, they may be grouped as follows:

1. *The well-drained soils.*—Well-drained soils are found in the area extending north and south through the central part of the Florida peninsula as far south as Highlands County, with lateral extensions into Pasco, Hillsborough, and Pinellas counties. The main area from Haines City to Lake Placid is often spoken of as the Ridge. Isolated areas of well-drained soils are found along both coasts and even on some of the coastal islands.

These soils belong principally to the following series: Astatula (formerly Lakeland and, earlier, Norfolk), Blanton, Eustis, and Orlando in the non-phosphatic group; Ft. Meade, Gainesville, and Arredondo in the phosphatic group. These are the soils most extensively used for citrus culture as far north as winter temperatures permit.

Astatula fine sand is the predominant soil for citrus plantings throughout the greater portion of the citrus area of the state. Its general characteristics are: usually sandy throughout the profile, although it may contain a small amount of clay at varying depths; surface 3–4 inches of brownish-gray, fine sand, with the subsoil of yellow fine sand; water table very low most of the time; although during the rainy season it may temporarily be high; soil reaction (acidity) under natural conditions ranging from pH 4.8 to 5.4, but under good cultural practices in citrus groves usually maintained at pH 5.0–6.0 or even slightly higher; organic matter very low (1–2 per cent), producing a low exchange capacity for mineral nutrient elements.

Eustis, Orlando, Gainesville, and Arredondo series present slightly better ability to hold water and nutrients, and so produce better tree growth than Astatula.

The main advantages possessed by the well-drained soils are good water drainage, good air drainage, and good depth for root development. Trouble is seldom encountered from waterlogging of these soils, although drainage may sometimes be slow because of compacted soil layers; usually drainage is only too good, resulting in deficient supplies of soil moisture. Although these soils are relatively frost-free because of their elevation, "frost pockets" frequently are present in the low areas between hills and must be avoided in planting. Soil uniformity, lack of hardpan, and generally low water table allow for deep, well-ramified root systems. Such an extensive rooting habit enables the tree to overcome to a considerable extent the handicaps of the low ability of these soils to hold water and nutrient elements. Deep, well-branched root systems enable trees to attain standard size and enjoy long life.

The disadvantages of these well-drained soils lie in their initial low level of mineral nutrients and organic matter and their small ability to hold water and the mineral elements applied as fertilizers. Excessive leaching through the centuries, together with the low base-exchange capacity, is responsible for the low nutrient level in the soil. Exchange capacity is based on the relative amounts of clay and organic matter in the soil, and both are low in these soils. The low content of the same substances reduces the ability of these soils to hold water, too. When clay occurs close to the surface, the groves respond far better to fertilizer applications than when it is present deep down or not at all. Any program of fertilization must take into account this low exchange capacity. As above suggested, development of a root system which occupies the soil very thoroughly to a good depth can overcome these disadvantages to a remarkable degree.

The water-holding ability of a soil is a function of its texture (the size of the particles composing it) and its structure. Sandy soils have large particles which hold water with very low forces. The single-grain structure of sands adds nothing to this low water-holding ability. Thus the capacity of these soils to hold water is slight for any given volume of soil, and a rainfall deficiency at any time of year is soon felt by the tree because this low reserve of water is soon exhausted. But a deep sandy soil has a great total volume of water, even though the soil moisture is low in any portion of it, and trees with a deep, well-branched root system can draw on this relatively large total reserve. Thus groves on the deep, well-drained soils with low water-holding ability may still withstand a prolonged drought better than groves on shallow soils of high water-holding ability.

2. *The poorly drained soils of the coastal areas.*—The low hammock

lands are found principally along the east and west coasts, although they also occur in isolated areas in the central portion of the state. They are represented by the Parkwood, Manatee, and Bradenton series. In general these soils are underlaid by marl (very finely divided calcium carbonate). The reaction of the surface soil may be as low as pH 5.0 while that of the subsoil increases to pH 8.5. The organic matter content is high, ranging from 3 to 8 per cent. Usually the water table is close to the surface, and in the natural state of these soils it may often be above the soil surface. They must be drained and bedded before they can be used for citrus planting.

The principal advantages of these soils are their high natural nutrient supply, their high base-exchange capacity, and their good water supply. The high exchange capacity permits better utilization by the trees of the nutrient elements present or added in fertilizer to the acid, surface layer. This is reflected in the great vigor of growth usually exhibited by trees in the early years of the grove life.

These soils also have several disadvantages for use in citrus groves, among them: poor air drainage, poor water drainage, shallow depth with small possible root volume, and alkalinity of the subsoil. Although the soils of this type used for citrus planting are located geographically where frost hazard is relatively small, they are by no means free from frost and their low elevation does not allow for compensating air drainage. Ditching systems are absolutely necessary for these soils and will take care of normal problems of water drainage. In times of very heavy rainfall, however, such as may occur in the hurricane season, the ability of the ditching system to remove excess water rapidly may be much exceeded. Pumping equipment may be needed to supplement the ditches, for excess water must not be allowed to stand long on the grove; and costs of removing this water add greatly to total production costs. Political subdivisions, set up as "drainage districts" supported by taxes, have long been established on the east coast to handle the area drainage into which water from individual groves is pumped.

Shallow depth to water table means that roots can only be formed to a shallow depth also, and thus only a small volume of soil is available for storing available water. In periods of drought the moisture supply to root depth is quickly exhausted, so that these soils are "droughty" in spite of their high water table. Irrigation water must be applied in small amounts frequently, which makes its application more expensive than on the deep, well-drained soils.

The alkalinity of the subsoil, often pH 8.5, renders unavailable to the tree many of the mineral elements of the micronutrient group—copper, zinc, iron, manganese, and boron—which decrease in availability as the soil reaction changes from acid to basic. Potassium and magnesium must be present in greater amounts before the tree can get an adequate supply because of the effect of the high calcium content of the subsoil. This situation has made it necessary for the grower to supply most of the above micronutrients as sprays—the nutritional spray program—because they are more readily absorbed through the leaves than through the roots in these soils. In spite of the alkaline subsoil, the grower must often make applications of lime to correct a tendency for the surface soil to become too acid because of continued applications of sulfur for pest control.

As the result of the shallow root system and erratic soil moisture supply, sometimes accentuated by a hardpan layer, citrus trees tend to be somewhat stunted in size and short-lived on these soils, and to produce low yields. Experimental work on the east coast has indicated that the root systems of citrus trees are often further restricted by temporary high water tables which prevent the efficient use of all of the already shallow theoretical depth for rooting. Practices which lower the water table and increase the depth available for root development on these soils will usually return more from the investment than any other practices which the grower may undertake.

3. *The flatwood and marsh soils of central and south Florida.*—The flatwood soils and associated swamp soils are found in any area of the citrus belt as soon as one leaves the elevated central Ridge. For years these soils were used for open range of cattle and hogs; then improved pastures were developed. They were considered quite unsuited for citrus plantings in general. Manatee and Parkwood and their associated soils were planted to citrus on the east coast, as noted above, and Leon and Immokalee of the flatwoods were used in a small way for citrus plantings.

However, as urban and suburban populations have increased during the last two decades, a premium has been placed on the well-drained soil with expansion of existing communities and development of new population centers and megapolis-type sprawls. There appears to be no end in sight to the influx of new residents into the state. This has forced the new plantings of citrus onto a whole range of soil types, acid and alkaline in reaction and with varying degrees of drainage, for which there is no long-term commercial experience. With the establishment of drainage districts (see above) or large-scale area drainage by citrus corporations, and the installation of ditches, dikes, and low-pressure pumps of high volume, it is possible to provide artificially a drainage not present naturally.

Groves are planted on 1-, 2-, or multiple-row beds; spodic or hardpan layers are broken up by deep plowing or use of a dragline; limestone is incorporated in those soils with acid reaction. Water-table wells, open-ended pipes seated on pervious gravel, are placed across the beds in various locations within the grove to determine the action of subsurface water. Aside from problems with high water levels, accumulation of sulphides due to aeration impairment has sometimes been serious. Extreme depths of white "sugar" sands have impeded good rooting, and cold damages, due to lack of air drainage resulting from uniformly low relief of the ground, may sometimes be suffered.

4. *The oolitic limestone soils.*—One soil series in particular, Rockdale, occurs in Dade County and is extensively used for groves of various citrus fruits, especially limes, and avocados and mangos. A very shallow layer of clay or sand overlies a very porous limestone formed by the cementing together of the shells of tiny marine organisms. To provide adequate depth for root development this oolite must be broken up by dynamite or by heavy scarifying plows. Although the surface is not many feet above the water table in winter, the roots can penetrate only a few feet at most and trees easily suffer from drought in the dry season. In summer the water may be above the surface for some time in periods of very heavy rainfall, for while the soil is porous, drainage is slow because all of the land is near sea level.

In the early period of citrus growing in Florida, the grower could select land from observations on the native flora. Beginning about 1930 this was no longer feasible in most areas because the native vegetation had been cut off, and the grower has used soil types and natural drainage as guides in selecting grove sites. Since about 1950 there has not been enough land with naturally good drainage to take care of expanding demand, and the grower must now expect to incur greater expense for land preparation than formerly in developing new plantings. Utilization of artificially-drained soil means greater initial capital outlay, greater operating expense to maintain good drainage, and greater vigilance on the part of the grower to guard against high water table and cold.

Factors of Importance in Choosing Soils for Citrus Planting

Several factors should be considered in appraising a given soil for suitability for citrus planting. Two of them, water drainage and air drainage, are of major and equal importance, and a third one, depth for rooting, is closely related to the first two.

No single factor is more important than water drainage, for any accumulation of free water in the root zone results in poor aeration of the absorbing rootlets, causing impaired ability to absorb water and nutrients. If long continued, this condition will cause progressive injury and death of the roots, leading to decreased vigor and productiveness of the top. Such trees are short-lived. Water damage may be *chronic,* affecting the tree throughout its life, or *acute,* affecting the tree only for a brief period due to temporary lack of water drainage. Poor drainage also reduces the resistance of the roots to infection by foot rot and other soil-borne diseases.

Good air drainage is increasingly important as winter minima become lower, that is, as one goes northward up the peninsula or inward from the Atlantic seaboard. Low-lying soils are colder than elevated soils because there is no place for cold air to drain away to. Cold pockets—closed valleys in the hilly areas—are always hazardous planting sites, because on frosty nights cold air flows down into them from the adjacent elevated areas, forcing the warm air up above tree height. After a severe cold wave there is usually visible a bench line around the sides of such a pocket, below which trees are dead and above which they are alive.

FIG. 4. View of Apshawa Groves, Lake County, looking south from headquarters building.

Courtesy George R. Bailey, Jr.

Deep, well-drained soils with no impervious layers allow good depth for root development, which tends to produce trees of standard size, heavy fruit production, and long life. Conversely, shallow soils, whether shallow because of hardpan layers or high water tables, limit the size, bearing capacity, and life of citrus trees.

If a soil satisfies the above three criteria, then attention may well be given to such factors of secondary importance as water-holding capacity, nutrient supply, and soil reaction. To the extent that a given soil ranks high in regard to favorable levels of these factors, it will permit lower costs of operation by the grower to achieve the same degree of vigor and productiveness.

Water-holding capacity is influenced by the texture (particle size) of the soil and especially by the content of colloidal material such as clay and organic matter. In addition, the volume (depth) of soil through which roots can ramify plays an important role. Each soil has a characteristic *field capacity* (the percentage of soil moisture remaining after saturation and free drainage) and *wilting percentage* (the amount of moisture remaining after plants have reduced the water content until they have become permanently wilted). The difference between these two percentages represents the *available water,* that which can readily be used by the plant. The higher the percentage of available water, the greater the supply in each foot of soil for tree use; and the greater the depth of the soil, the greater the total reservoir of water to sustain the tree in droughts. The amount of available water in a given depth of soil can be computed in inches of water, equivalent to inches of rainfall, and as such gives a valuable clue to whether or not there is need for irrigation during periods of extended drought.

The nutrient supply in most of the soils used for citrus groves in Florida does not vary greatly in the natural state because these sandy soils have been leached by high rainfall for millenia and all have a very low level. However, soils of relatively high organic content have a somewhat greater original supply of nutrients than those of very low organic content. More important is the fact that soils of high water-holding capacity—those with relatively high content of clay and organic colloids—also hold mineral nutrients applied as fertilizers against loss by leaching much more effectively than soils low in colloids.

Florida soils are classified as acid or alkaline in reaction, a condition customarily expressed on a scale of pH units where 7.0 represents the neutral reaction, values below 7 indicate increasing acidity, and values above 7 increasing alkalinity. The most favorable reaction for citrus trees on sandy soils is pH 5.5 to 6.5, or slightly acid conditions. Most sandy soils are naturally more acid than this and require periodic applications of lime to maintain the desired reaction. The alkaline soils of Florida cannot be acidified to reach this desired pH level, and require certain specialized management practices in compensation, notably in the application of minor elements made unavailable by the alkaline reaction.

Areas of Citrus Production in Florida

The first citrus plantings were made in the vicinity of St. Augustine around 1565 at the time of the establishment of that settlement. As the peninsula became populated, citrus seedlings were spread throughout the state. Citrus growing expanded from home orchards of seedlings and budlings to commercial acreages of budded trees. The citrus industry of Florida, in 1890, comprised approximately 114,800 acres with the 10 leading counties in the following order: Orange, Alachua, Volusia, Marion, Lake, Putnam, Hillsborough, Pasco, Brevard, and Polk. Alachua County accounted for about one-third of the citrus production of 1889–90. Production increased to an estimated 6,000,000 production for 1894–95. Then, in December of 1894, a freeze occurred that defoliated the trees; this was followed by a second freeze on February 6, 1895, which killed most of the trees. Acreage was reduced to 48,200 acres in 1895, of which 97 per cent was of non-bearing trees. Orange, Lake, and Brevard counties accounted for over one-half of the acreage remaining after the freezes. Polk County, which later became the leading county, was tenth in the county list; the growers had yet to plant the large expanse of rolling, sandy soils. A cold wave in February of 1899 brought some of the lowest temperatures experienced in the state and produced excessive damages to the industry. These freezes proved the necessity of moving the industry southward to locations less vulnerable to low winter temperatures.

During the Florida land boom of 1922–28, large-scale plantings were made in the central parts of the state and along the east coast and west coast. These set the pattern of distribution of the industry (as was discussed in the first edition) for many years. Much of this boom-developed acreage received little attention, and some was forced into a semi-abandoned condition during the long depression of the 1930's. However, this depressed period produced leaders in research and production to whom credit must be given for the continued health of the industry.

Further plantings were made, and abandoned acreages were brought back to productive condition, with the resumption of demand for citrus during the 1940's. By 1955 there was 522,000 acres of citrus with the leading counties being Polk, Lake, Orange, Hillsborough, St. Lucie, Pasco, Indian River, Highlands, Brevard, and Volusia.

However, the great influx of new residents from the north beginning in the 1950's has exerted a strong influence on the distribution and dominance of the industry. These new settlers began to occupy the best-drained areas along the coasts and in the central portions of the state; counties such as Orange and Pinellas began to lose many acres of citrus to the urbanites; land values and taxes began to increase to the economic detriment of citrus production. Citrus plantings were forced into less desirable locations. An attempt was made to expand citrus plantings along the upper west coast but the winters of 1957–58 and 1962–63, with their severe freezes, aborted this move.

Therefore, the second great move has been to the south onto soils which had earlier been avoided because of drainage problems. This move started in earnest following the 1962 freeze and reached a climax in 1969, at which time approximately half of the citrus acreage (and an even greater percentage of citrus trees due to closer planting schemes) was land of varying degrees of impeded drainage. These mineral soils of the flatwoods and marshes of central and south Florida presented to the growers and research workers many new problems which were quite distinct from those of the older areas for which solutions had been found through long years of research and commercial experience. By 1969 the better grades of even the poorly drained soils had been occupied, slowing the tempo of plantings. Furthermore, the federal income tax reforms of 1969 forced the capitalization of all expenditures during the first four years of a new planting, making such plantings far less desirable to developers and speculators. Although new plantings will continue to be made, it currently appears that these will be on a reduced scale; it will be difficult to replace the acreage lost to urban developments.

During this most recent expansion of the industry, large acreages of citrus were planted in the flatwood areas of the Indian River counties of Brevard, Indian River, St. Lucie, Martin, and Palm Beach, and in the flatwood and swamp soils of counties in the southern part of the peninsula. By the late 1960's, the total acreage of citrus had reached almost 1,000,000. However, in the early 1970's, the new plantings did not keep pace with removals so that in 1971 there were approximately 877,000 acres with the principal counties being: Polk, Lake, St. Lucie, Orange, Indian River, Hardee, Hillsborough, Martin, Highlands, and Pasco.

The general citrus-growing portion of Florida lies within the limits bounded on the north by a line through McIntosh and Palatka with small extensions northward along the St. Johns River and on the south by the Everglades with extensions on the east coast to Homestead and on the west coast to Naples. Beyond the northern limits, into every county in the state, will be found a few dooryard citrus trees given protection against the cold, northwest winds of winter. The principal area of lime production is along the lower east coast and southern part of the Ridge; that of lemons, the southern third of the peninsula and particularly Palm Beach, Martin, and Hillsborough counties.

Approximately 32 of the 67 counties grow and ship citrus fruits in commercial quantities and all of them lie within the area described above. To the north, winter temperatures limit commercial growing of citrus trees while counties at the southern limit frequently find either tourist competition or extremely poor soil drainage disadvantageous. The citrus crop is of major economic importance in many of these 32 counties and, by far, the leading source of income in some. Polk, Lake, and Orange counties, in the center of the state, were for years the three leading producers. However, with increased plantings beginning in 1959 in the western portions of the counties, St. Lucie and Indian River have come into prominence. Currently, St. Lucie has nudged Orange County out of third place while Indian River has advanced to fifth place. Tables 4 and 5 show how counties compare in acreage and production of citrus.

Environmental conditions, adaptation of stocks and scions, general cultural practices, and marketing opportunities vary considerably in different parts of the commercial citrus belt. The prospective investor in grove properties should study these factors in the area of his interest. He may obtain help from established growers and production managers, county extension directors and specialists, and representatives of allied industries and marketing agencies. Long experience has shown the value of consideration of land values and taxes, availability of production and market services, and general levels of production costs and returns. Studies of climatic, soil, and biotic conditions, stock and scion selections, and general conditions of grove properties with respect to tree size for age, uniformity of tree size, regularity of plantings, and incidences of inherent problems such as nematodes, viruses, frosts and freezes, soil drainage, etc., will be most valuable for the person who will be intimately involved in the industry and especially for the person in a managerial position. When soil drainage presents a problem, one must consider area drainage (through drainage district or otherwise) in order to understand the drainage of a particular site or location. There are areas of better soils that could likely produce citrus successfully if provision for heating is considered.

In the first edition 6 areas of production were discussed: the Indian River, the lower east coast, the lower west coast, the upper west coast, the north central, and the south central. These had developed from 1920 to 1960 through the community of interest of growers and the similarity of problems in each of these areas. The cohesiveness of these areas has changed considerably in the last decade with the

TABLE 4

CITRUS-PRODUCING COUNTIES OF FLORIDA
(Acreage by county and type of fruit as of December, 1971)

County	Oranges	Grapefruit	Tangerines and Hybrids*	Other	Total
Polk	108,449	25,455	9,528	721	144,153
Lake	101,652	12,374	17,335	1,313	132,674
St. Lucie	39,647	25,746	8,129	300	73,822
Orange	49,121	2,714	8,282	450	60,567
Indian River	24,697	24,206	3,007	142	52,052
Hardee	43,025	839	1,815	307	45,986
Hillsborough	36,822	2,504	2,444	1,142	42,912
Martin	30,180	5,261	1,846	4,071	41,358
Highlands	28,740	4,432	4,043	550	37,765
Pasco	32,937	1,399	2,265	184	36,785
DeSoto	26,186	1,105	1,550	233	29,074
Hendry	17,098	2,485	2,155	946	22,684
Brevard	14,162	3,212	1,066	64	18,504
Osceola	14,857	929	1,765	36	17,587
Palm Beach	9,482	3,618	3,405	983	17,488
Manatee	12,171	1,972	1,526	152	15,821
Marion	10,680	395	508	201	11,784
Volusia	9,508	683	1,273	218	11,682
Seminole	8,735	304	1,822	108	10,969
Lee	6,419	499	260	112	7,290
Hernando	5,983	146	769	100	6,998
Charlotte	5,691	286	446	217	6,640
Collier	4,613	335	215	65	5,228
Pinellas	3,013	1,644	237	42	4,936
Broward	4,002	538	182	50	4,772
Dade				4,403	4,403
Okeechobee	2,752	609	304	11	3,676
Putnam	2,803	51	455	131	3,440
Sumter	1,728	7	15	21	1,771
Glades	1,405	75	79	80	1,639
Sarasota	1,120	285	115	19	1,539
Citrus	1,355	32	127	22	1,536
Other	385	2	74	23	484
Total	659,418	124,142	77,042	17,417	878,019

Source: Florida Agricultural Statistics, Citrus Summary, 1972

*Including Temples, tangelos, tangerines, and Murcotts marketed as specialty citrus fruits.

change in the distribution of acreage and the expansion of organizations with plantings in various parts of the state. The least change has occurred in the Indian River and south central areas.

The Indian River Citrus Area is defined by state law (Acts of 1941) which specifies that the name may be used only for fruit grown on

TABLE 5
CITRUS CROPS IN FLORIDA BY COUNTIES FOR 1971–72*
(in 1,000 boxes)

County	Oranges	Grapefruit	Tangerines and Hybrids**	Total
Polk	30,764	13,956	1,819	46,539
Lake	20,471	5,860	3,584	29,915
St. Lucie	7,987	7,445	1,430	16,862
Orange	12,940	1,422	1,933	16,295
Indian River	4,219	7,225	577	12,021
Highlands	7,291	2,327	590	10,208
Hardee	8,644	338	213	9,195
Hillsborough	5,588	865	395	6,848
Brevard	4.252	1,386	220	5,858
Pasco	5,022	475	327	5,824
Martin	4,324	878	283	5,485
DeSoto	3,555	433	160	4,148
Osceola	2,574	375	323	3,272
Volusia	2,531	374	251	3,156
Marion	2,552	237	91	2,880
Hendry	2,272	278	205	2,755
Palm Beach	1,432	755	506	2,693
Seminole	2,013	184	338	2,535
Manatee	1,499	716	277	2,492
Pinellas	783	687	49	1,519
Broward	1,005	199	31	1,235
Hernando	839	57	112	1,008
Charlotte	811	82	58	951
Putnam	835	34	71	940
Lee	622	161	40	823
Collier	660	42	30	732
Okeechobee	387	64	25	476
Sumter	348	5	4	357
Sarasota	205	112	28	345
Citrus	236	24	20	280
Glades	215	2	1	218
Other	124	2	9	135
Total	137,000	47,000	14,000	198,000

Source: Florida Agricultural Statistics, Citrus Summary, 1972

*Does not include lemons and limes grown in Dade, Martin, Palm Beach, Hardee, Hillsborough, and Highlands counties.

**Including Temples, tangelos, tangerines, and Murcotts marketed as specialty citrus fruits.

land adjacent to the Indian River, and lying totally within the area described by law, along the east coast. It includes Brevard, Indian River, St. Lucie, Martin, the southeastern part of Volusia, and the northern part of Palm Beach counties. Since fruit grown in this area, especially grapefruit, has a recognized superiority on the market, as evidenced by demand and prices, it is accorded separate regulations under the Marketing Agreement. Fruit grown in all other sections of the state is designated as "interior fruit" for purposes of marketing in the fresh-fruit channels of trade. The south central area, comprising the counties of Polk and Highlands, consists of large expanses of Astatula (Lakeland) fine sand of undulating nature with the various types of oranges, grapefruit, and tangerines budded, in the main, on 'Rough' lemon rootstock, the stock now under a cloud because of a new disease which causes trees to decline.

On the accompanying map, four citrus-producing areas are delineated as defined by the Florida Crop and Livestock Reporting Service for recording acreage and production of citrus each year. The East Coast Area includes the counties of Brevard, Broward, Dade, Indian River, Martin, Palm Beach, and St. Lucie. Alachua, Lake, Marion, Orange, Osceola, Putnam, Seminole, Sumter, and Volusia counties comprise the Upper Interior Area. In the Lower Interior Area are Charlotte, Collier, DeSoto, Glades, Hardee, Hendry, Highlands, Lee, Okeechobee, and Polk counties. The West Coast Area includes Citrus, Hernando, Hillsborough, Manatee, Pasco, Pinellas, and Sarasota counties.

County lines, as political boundaries, leave much to be desired in describing areas of similar climatic, soil, and biotic environments. One area grades almost imperceptibly into another and within any area there will be islands of differences. Note that the East Coast Area includes all of the Indian River Citrus Area except for southeastern Volusia County. The East Coast, Upper Interior, and West Coast areas present few extreme differences within their boundaries, but the Lower Interior Area includes the well-drained counties of Polk and Highlands among counties which are generally poorly drained. The small amount of fruit grown north of the Upper Interior Area, in the counties of St. Johns, Duval, Clay, and Flagler, is included in this area. The actual citrus-producing belt of Florida, by legal designation, extends north and west to the Suwannee River.

The study of the area is a most valuable exercise. However, each particular site or location within an area presents its unique set of conditions, ranging from best to poorest of area, which determine its level of value to the investor. Once satisfied with the general con-

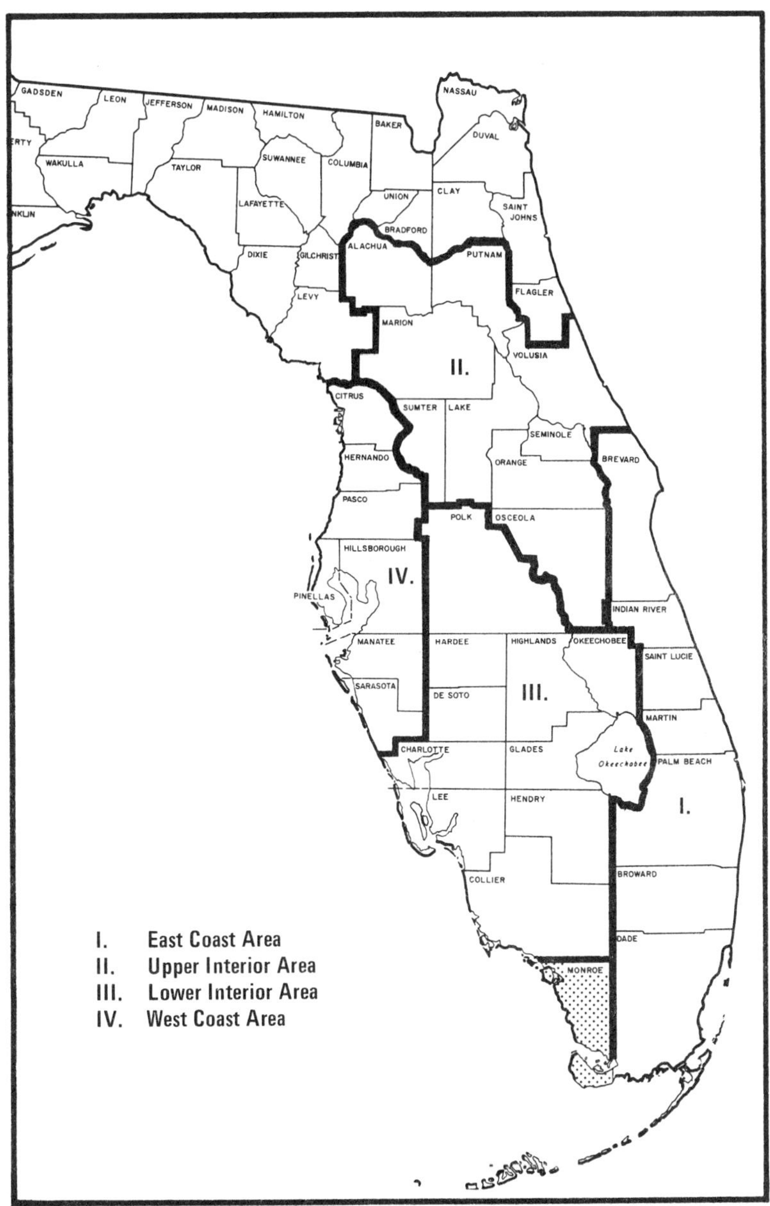

Courtesy R. Knudsen

ditions of the area, the investor will seek the best of these items: natural cold protection; soil depth and drainage; land or grove costs (original, preparation, and continuing costs especially when concerned with soil drainage); and tax structure. Small differences with respect to conveniences may be considered as plus or minus factors but should never take precedence over the first four in the consideration of the site. The purchaser must have in mind his long-term objectives as an investment for citrus growing, or for land speculative possibilities, or both, which will greatly assist in making a proper choice.

5. Planting Citrus Trees

Selection of the Grove Site

THE first step toward planting a citrus grove is the selection of a suitable location. In view of the fact that satisfactory planting sites for citrus in Florida are becoming less available, the prospective citrus grower may often be concerned with the purchase of an existing grove. It should, therefore, be stressed that the following discussion applies, with similar effect, to the site already occupied by trees.

Two sets of factors are of dominant importance to the future economic well-being of the enterprise. The first of these relates to land values and the tax base. While the investor is interested in the potential increase in land value in future years, he will restrict his search to areas which are still predominantly agricultural to avail himself of the community of interests of other citrus growers. Urban expansion results in zoning changes and increased tax burdens for services which often render no advantage to the citrus enterprise. The matters of the total investment in view of increasing property values and taxation require the investor's keen judgment in order to realize optimum competitive economic returns.

The second set of factors are those of horticultural importance to the growing of the crop. Predominant are the necessities for natural protection against low winter temperatures and good soil drainage. Depth for rooting is essentially related to soil drainage in the Florida citrus belt. Other considerations are discussed which will help to maintain the soundness of the enterprise.

Cold Protection

Natural protection from cold will vary from area to area, from section to section within an area, and from one location to another within a section. Each of the citrus areas of the state has its own general range of climatic conditions. Within any area slight differences in latitude or in proximity to ocean or gulf cause differences of a few degrees in winter minimal temperatures. At the same latitude or distance from the coast there will usually be found variations between general locations caused by differences in elevation, vegetation, bodies of water, etc. Old growers in any section of the state can point out situations which are, in general, colder than others, even though they appear to be similarly located.

The grove site itself is of particular concern to the prospective grower. Elevation is of prime importance, because a site higher than nearby land will allow cold air to drain away on frosty nights. So far as possible a site should be avoided which is in one of the natural air channels followed by cold air masses bringing freezes to the state. The moderating influence of bodies of water on air temperatures has often been noted. These bodies must be large and deep to exert any marked effect. Usually the southeast side of a large lake is the warmest in a freeze because the coldest winds come from the northwest and are warmed somewhat in passing over the water. The Gulf of Mexico is able to influence temperatures appreciably, but Florida has few lakes large and deep enough to ameliorate the cold air blowing over them. Small lakes, especially if they are shallow, are of no value in warming masses of cold air, but they may still be an asset for cold-protection in that they provide a fairly large, level area at low elevation for cold air to drain down to from any surrounding uplands. The compass direction of these lakes from the grove site is unimportant. Grove sites in an area where many groves already exist will receive some cold-protection from these established plantings, which may function as windbreaks or may warm cold winds somewhat by artificial heating, to the benefit of any grove site to leeward. An isolated grove in thin, open pine woods is at the mercy of every freeze which descends into Florida.

Appraisal of the desirability of a site can be assisted by careful observations in old groves nearby. Regularity of stand, uniformity of tree size, and presence of frost cankers should be investigated. In areas of irregular topography, cold spots in old groves will be devoid of trees or will have trees showing poor condition. Even where there are not marked differences in topography, the presence of many vacant places in groves, or of many replanted trees of different ages, may indicate cold-trouble. Other possible causes for such irregularity should be carefully checked, to be sure whether or not cold is responsible. The grove with satisfactory site is one with uniform height and spread of the trees. Wide fluctuation in these dimensions from one tree to another indicates that some undesirable conditions exist. If low temperatures have influenced these conditions, further study will usually demonstrate a relationship to topographical features. It may sometimes seem that evidences of freeze damage are sufficiently slight to be discounted, but if they can be recognized at all they should be duly considered. The average size for their age of trees of any particular combination of scion and stock, for a given set of soil conditions, is an excellent guide to the possible future productiveness of the grove. If trees are even slightly below normal size because of cold-injury, their yields will also be slightly below normal—and profits will be also.

Freezes often produce small areas of dead tissues, known as "frost cankers," on the bark of scaffold branches and trunks. The surrounding living tissues continue to grow and form an advancing front of callus around the edge of the canker. The dead tissues gradually slough off, leaving a depressed area in the bark which remains for some years as mute evidence of past cold-damage. There is no way to get rid of them except by cutting off the whole limb, until after many years they may be obliterated by natural growth of bark.

The care exercised in the choice of a grove site for natural cold-protection is reflected in the later cost of production of the grove. Even at best, "firing" operations add greatly to costs of production and to the administrative load of the operator. If grove-heating is not practiced, then there are decreased yields from cold and these again increase the production cost per box of fruit. A site subject to cold-injury, no matter in what degree, sets the range of costs per box within which the management can operate.

Proper selection of stock and scion varieties with respect to their cold-hardiness can often render a cold site relatively more profitable, but cannot overcome entirely the basic handicap of a poorly chosen site. Likewise it is possible to develop operating programs which can reduce the hazard of cold, but these cannot be expected to compensate completely for the original lack of judgment in site selection. Too much emphasis cannot be placed on the permanent nature of the advantage or handicap for profitable grove operation involved in the choice of the grove site.

Soil Drainage

The greater the depth of soil for root development, the better the response of the trees to all production practices: trees will be larger, yields will be greater, production costs per box will be lower, and productive life of the trees will be longer.

The soil series, as explained previously, will give considerable clue to drainage conditions and depth for rooting. The characteristics of the type should be understood as thoroughly as possible. Even if the name of the soil type is known—and usually the County Extension Director can identify it—a careful appraisal of the particular site should be made. If the grower does not know the soil type, it is more than ever important that he make a thorough investigation of the nature of the drainage, the presence or absence of a hardpan layer, and any other textural or structural characters of the soil which will have an influence on either drainage of water or penetration of roots.

No citrus tree can be grown successfully where the whole of the possible soil volume for root development is occupied by water. Even if the water table is very close to the surface only for relatively brief periods each year, that location will usually have to be ruled out. Certainly the type of water relations characteristic of hardpan (spodic layer) soils—wet in summer, dry in winter—will be reflected very unfavorably in tree responses and production costs. However, roots will penetrate the pan layer if adequate drainage below the pan is provided. The pan can be broken up by deep plowing or dragline operations.

Most citrus investigators consider that the rooting depth of the soil, to be highly satisfactory, should be from 4 to 6 feet; and a depth of 2 feet is considered as the minimum under most circumstances. If the depth of soil for root development is not satisfactory, then the site should be avoided unless it is evident that there are ways in which the depth can be increased. But it must be remembered that such ways of increasing rooting depth as ditching and bedding mean increased capital outlay, and interest on capital investment is part of production cost. Furthermore there will annually be increased expense for current operations in water removal, including drainage district assessments, and these add to the total cost of grove operation. Cost of production per box will be increased also because of reduced response of trees to the production program—smaller trees with lower yields, and shorter life of trees. Grove operations may still be profitable, but the prospective grove owner should be aware of the factors involved in the use of a poorly drained site.

After satisfying himself as to satisfactory conditions of cold-protection and water drainage, the grower should investigate such secondary factors as soil characters other than drainage, facilities for marketing the crop, the possibility of other operational problems, and the availability of labor.

Soil characters other than drainage include the presence of clay in the soil, the level of natural nutrient supply, the content of organic matter, and the soil reaction. If two sites are equally well drained, then the choice may well depend on the extent to which one of them excels in these respects.

The site should be such that advantage can be taken of several different types of marketing. These include packinghouses, canning and concentrating plants, and gift-box stands. Transportation facilities are no longer a concern and paved roads are common in all parts of the citrus belt, yet the cost of hauling fruit to the packinghouse or concentrating plant is still based on the distance it is hauled. When the grove site is quite distant from the point where fruit enters the channels of trade, the grower is under an economic handicap. The type of market for which fruit is grown will determine to no small extent the choice of varieties planted.

Among other operational problems which affect production costs, besides soil characteristics, is control of pests. Most of the insects and diseases affecting citrus trees are present throughout the citrus-growing area of the state, but one or more of them may be a more serious problem in one section than in another. Tristeza virus and spreading-decline nematodes spread naturally from one grove to another. Other things being equal, a grove site adjacent to a grove in which either of these diseases is present would be much less desirable than a site fairly distant from any known infestation.

Labor costs have increased in recent years faster than any other item in cost of production, and at the same time competition for the labor pools has become more intense. Producers of vegetables, urban developments, the hotel and resort trade, military installations, and new factories all compete with the citrus grower and have constantly expanded their labor needs too. Even though the labor supply for grove work may be plentiful during some of the year, competition which interferes with this supply during certain critical periods may be a limiting factor. Such operations as firing during a freeze, irrigating in a spring drought, or applying sprays for critically-timed pest control in summer cannot be delayed while additional labor is rounded up.

Much of the operation of a citrus grove calls for unskilled or semiskilled labor only, and in a high-priced labor market the citrus grower is at a disadvantage. Labor problems often make more demands on the time of the grove manager than any other. The solution of the problem for the citrus grower, as for the industrial plant, is to mechanize his operation so far as possible, so that fewer and more-skilled laborers can be utilized. Marked advances have been made in recent years in mechanization of grove operations such as fertilizing, spraying, and pruning, but there still remains the need for a small labor force of unskilled but dependable men. Further reduction is quite possible for grove-heating, irrigation, chemical control of weeds, and even for harvesting. So long as unskilled labor is a necessity, the grove operator in a community where competition has bid up the wage scale above that in other areas is at an economic disadvantage.

In summary, it must be remembered when selecting a grove site that citrus trees must occupy the same ground for many years to be profitable, and the crop must mature on the trees each year. The vegetable grower may lose a crop one year and recoup his loss the following year, but the citrus grower must count on a regular crop every year. To the extent that he loses an occasional crop from cold or wind, he is losing profits permanently, and there is a limit to how many such losses he can take and still remain solvent. The level of crop production is determined at planting time by the site selected and the stocks and scion varieties chosen for that site. The small, short-lived tree on shallow, poorly drained soil is never an economically productive unit.

Preparing the Grove Site for Planting

Having selected with care a suitable grove site, in accordance with the considerations just set forth, this site must be cleared of all trees and shrubs. All large pieces of roots, and as much of small roots as possible, should be removed from the upper foot or two of soil, and the ground then plowed and harrowed until it presents a smooth, level surface. If the grove site itself is on a slope and it is considered desirable to plant on contoured benches, their preparation must await decision on planting distances to be used.

Drainage should be considered at this point for any site in a poorly drained area. Surveys will determine the slope of the land for proper run-off of surface water. Beds for single, double, or multiple rows of trees with furrows to allow for drainage are constructed. Ditches between blocks of beds are required at intervals dependent upon the character of internal drainage of the soil. Means should be considered to allow rapid drainage during periods of heavy rains; at the same time, accessibility of water for irrigation during dry periods must be planned. These are important considerations prior to the planting operation; consultations with extension personnel, engineers, and horticulturists may result in the necessity of slightly higher preparation costs but will greatly increase the future profitability of the enterprise.

Choice of a Planting System

By planting system is meant the regular arrangement of the trees on the land. Formerly several different systems were in vogue, each with its loyal advocates, such as rectangular, hexagonal, triangular, and quincunx. As with other orchard fruits, this former diversity has resolved itself into almost exclusive use of the rectangular system for citrus trees. Some of the other systems permitted planting of more trees per acre for a given spacing, but they required cultivation and spraying to be done along diagonals through the grove, which was troublesome and confusing when varieties were set in rows parallel to the boundaries of the grove.

The rectangular system has the trees at the intersections of lines which cross at right angles, and the rectangles may be square or oblong. The system is easy to lay out, to understand, and to use in grove operations. Such groves can be cultivated in two directions easily, and varieties in them can be planted in compact blocks of rows, making it easy to keep track of them in spraying or harvesting. Planting on the square was formerly by far the more common practice, but today it is desired to have trees closer in one direction than the other. In this case care must be taken to have the wider middles the more easily accessible ones, for when the trees reach some size, they will have to be the middles used for nearly all work.

The hedgerow system of planting was adopted by some of the early growers of citrus in Florida. It permitted planting the largest number of permanent trees per acre in a satisfactory pattern for commercial grove operation; space between the rows allowed equipment movement and space between the trees in the row was kept at a minimum. While trees were small and not crowded, higher yields were possible. This system required extremely good fertilizer and irrigation practices, especially on sandy soils low in organic matter, to effect satisfactory results; however, the problem of crowding in later years forced the general abandonment of the hedgerow system. With the closer plantings of the square or rectangular system under the current hedging, topping, and lifting practices, there would appear to be a return to the hedgerow; however, the approach of management today is much more scientifically sound.

In the contour system the trees are planted in rows along lines of equal elevation or contours. Its use is limited to slopes steep enough for soil erosion to be a serious problem by conventional planting systems. If slopes are so steep that it is difficult to drive machinery along the contour lines, it may be necessary to make beds several feet wide for planting, so that machines can run on nearly level ground. Laying out contours requires surveying skill and takes much more time than laying out a rectangular system. A contour planting is harder to work and to understand, and the rows along contours will often not all have the same number of trees. The system is rarely seen in Florida and should never be attempted unless it seems absolutely necessary for erosion control. Other methods of controlling soil erosion should be given careful consideration before deciding to plant on contours.

Double planting of the various systems came into vogue in the 1940's to allow the best utilization of land while the trees were small during the early years of grove life. It was a means of setting twice as many trees as usual on one axis in a standard planting with the anticipation of removing the alternate trees as soon as they began to crowd. It did not constitute a distinct system but was simply a short-term expedient. In many cases, tree removal was delayed by first cutting back the alternate trees before finally removing them to allow room for the permanent ones. Quite fortunately, the hedging practices were developed in the 1950's; tree size was controlled to space, permitting many of these double-set groves to continue as viable plantings without tree removal.

Spacing of Trees

Regardless of the planting system used, it is necessary to decide how far apart trees are to be placed. Spacing was originally determined chiefly by the size to which the trees were expected to grow. Trees that are small by nature, or that would remain small because of poor soil conditions, would be spaced more closely than trees which naturally reach large size and are growing in favorable soil.

Under similar conditions grapefruit trees grow much larger than orange trees, which in turn exceed the size of lime trees. Average spacing in the row for the various kinds of citrus trees was: kumquats, 10 to 12 feet; satsumas, limes, and lemons, 15 to 20 feet; oranges and mandarins (except satsumas), 20 to 25 feet; and grapefruit, 30 to 35 feet. It was, however, desirable in commercial groves to have the work middles at least 25 feet in width in order to be able to use mechanized equipment such as sprayers, fertilizer distributors, and picking trucks.

The early plantings developed into canopy groves. The bottom branches died out as a result of shading and the fruit crop was borne by a canopy of branches overhead. Fruit was difficult to harvest and yields were limited by the scant amount of fruiting wood.

A larger number of trees per acre does not necessarily produce a larger yield per acre; but planting too few trees to utilize soil and light efficiently is certainly uneconomical use of space. During the Florida boom of the 1920's common spacings, on the square system, were 25 by 25 feet and 30 by 30 feet, giving, respectively, about 70 and 48 trees per acre. On deep, well-drained soils, 25 by 25 became too close eventually for ease of operations even for oranges; the same number of trees per acre was possible by use of 20 by 30 spacing to get a better work middle by crowding the trees slightly in the row. There was a suggestion in the early 1940's that the best spacing for grapefruit trees would be 35 by 35 feet, which would allow only 35.5 trees per acre.

The citrus grower, as well as the grower of other fruits, has long been troubled by this problem of how to use the grove space more fully while the trees are young without having them too crowded at maturity. The solution used during the 1940's was double planting (see above). In theory this close initial planting and later removal of alternate trees after a few crops was a good solution. It was feasible to transplant the temporary trees to a new grove site if further plantings were being made. Growers, however, postponed as long as possible the removal of the non-permanent trees because they were producing valuable crops; usually thinning operations were put off until the permanent trees had suffered considerable damage from crowding. A gradual thinning gave some promise. Alternate trees were pruned sufficiently to leave a 2-foot clear space between them and their permanent neighbors. Each year these same alternates could be further headed back so that they never shaded the permanent trees. After 12 to 15 years in a 15 by 25 planting they had been reduced to skeletons and could be removed entirely, leaving the unchecked permanent trees spaced 30 by 25 feet. Too often growers refused to cut into any good tree and resisted the efforts of production managers to do so. Therefore, it became apparent that it was better to sacrifice a little income during the early years of grove life, because trees were widely spaced, than to gain the extra return from close spacing at the expense of misshapen trees of poor bearing habit in the mature grove, because the interplanted trees were left too long in double plantings.

After much fumbling, the brilliant idea of hedging began as a practice in the 1950's. It began as a remedial operation simply for the safe operation of equipment through the groves. It proved, however, to be a means for controlling tree size to space with maximum production per acre for closely spaced groves, with many advantages and no disadvantages, as discussed in Chapter 10. Today, tree spacing and hedging are considered to be maintenance operations. Current plantings of 12½ by 25 feet (139 trees per acre) are feasible and productive.

Staking the Tree Locations

After planting system and spacing have been decided upon, the tree locations may be marked by stakes. Boundary rows are first laid out, and then parallel rows established across the field. Once the tree locations are staked on two sides of a field by careful measurement, and perhaps a central row similarly measured and staked, the rest of the locations can be quickly sighted in and staked. Sometimes the tree positions are marked initially by crosshatching the ground by means of an outrider on a tractor, with the trees set where the lines intersect. Other methods have been used, any one of which is satisfactory provided the tree locations are in ordered rows.

The Transplanting Operation

Selection of Trees in the Nursery

The question of the most desirable size and age of tree to plant still remains for consideration even though the decision as to stock and scion variety has long since been made. Trees offered by Florida citrus nurseries are usually 1-year buds on 3-year stocks; i.e., the stock was a 2-year-old seedling when budded and the budling is a year old at digging. Such a tree is a 1-year-old by nursery convention, which disregards age of the stock. A desirable 1-year tree should "caliper" (i.e., have a diameter at a point a couple of inches above the bud union) at least ½ inch, and ⅝ inch is better. The pruning cut made in removing the stock above the bud should have healed over and be hardly noticeable, and the trunk should be straight. Such a 1-year tree is probably the best one to buy, since it shows good nursery care and gives evidence of having made vigorous growth as a budling.

A nursery may dig these better trees and leave the smaller ones to grow another year, at which time they are 2-year trees. There is every reason to believe that the original slow growth was inherent and that these trees will continue to grow more slowly than the others after transplanting. They are to be avoided. However, nurseries may undertake custom budding with a contract to produce 2-year-old trees for grove setting. Such trees should be spaced more widely in the nursery row than trees to be sold at 1 year, or the necessary root pruning in digging may reduce the value of the greater size of these trees. The grower pays a higher price for these trees, but has a larger tree to set out and a shorter period to wait for bearing. Such trees, properly handled in the nursery, are superior to yearling trees but more costly in proportion to size. But trees left unsold in the nursery row for a year or two have to be pruned severely at transplanting and develop more slowly into grove trees, in spite of their apparent larger size, than vigorous 1-year trees.

Price is a fairly good indication of tree value if one is dealing with

FIG. 5. Large-scale citrus planting operations.
Courtesy Adams Citrus Nursery, Inc.

a reliable nurseryman, and one gets about what one pays for. Nursery trees of unusual vigor, say ¾-inch caliper for 1-year budlings, bring a higher price because they will grow faster in the grove than smaller trees. Trees which have been neglected in the nursery or have been improperly handled (especially if they have been dug with less than the minimum length taproot, 12 inches) are very costly trees in the long run though bought at bargain-counter prices. The cost of the trees is a rather small item in the total cost of bringing a grove into production, and this is a very poor place to economize. Current market prices for good nursery trees can easily be ascertained by reading the advertisements in citrus magazines, remembering that the price per tree in small lots must always be a little higher than it is in large quantities.

The only assurance which the grower has as to variety and stocks, so that he is certain to get what he thinks he is buying, is in the integrity and care of the nurseryman. Florida has many citrus nurseries which have established and are careful to maintain a good reputation for nursery trees of good quality and assured name. The transient nurseryman may do as good a job, but the grower does well to investigate carefully before buying from an unknown source.

Trees propagated by nurseries with plant material registered by the Citrus Budwood Registration Bureau of the Florida Department of Agriculture and Consumer Services can be obtained free of psorosis, the most damaging virus of the early citrus plantings in Florida. Freedom from certain other viruses, propagable disorders, and abnormal genetic characteristics are possible in many varieties. The cost of such trees is higher than it is for trees from unregistered stock due to the extra efforts of the nurserymen, but this additional cost has proved to be an excellent investment. Repayment will come through longer periods of prime-yield life of such high quality trees.

Handling and Delivery of Nursery Trees

The first step in preparing nursery trees for transplanting is to prune the top. Much of the root system will be lost in digging, and this decreased capacity for absorbing water requires a decrease in demand of water by the top.

The technique of digging nursery trees has already been given in Chapter 3. Covering the tops with a moisture-retaining material is especially important if the trees are transported some distance by truck before planting. If it is necessary to delay planting after trees reach the grove site, they may be held for several days by "heeling them in," that is, by making a shallow trench and placing the roots in it while the trunk is inclined at an angle of 60° or so from the vertical. The roots are covered with soil which must be kept moist. It is always better, however, to get trees into their permanent locations in the grove the day they are dug.

Some of the citrus cooperatives have been growing trees in bushel hampers for use in replacing nursery trees which fail to survive transplanting. The cost per tree is considerably greater than for the usual nursery trees, but there is no loss of roots or top when transplanting these hamper-grown trees. Container-grown trees require less maintenance during the first year. Furthermore, resetting can be done later in the spring than with bare-rooted transplants. Fresh nursery trees would be many weeks behind the rest of the planting in getting started, whereas these container-grown trees are fully as advanced in growth as the earlier planted majority, and thus a more uniform grove is possible.

Season of Planting

Citrus trees in Florida are planted mostly during the winter months (December-February) while the trees are most fully dormant. Roots are able to start regenerating an absorbing system before the buds push out into shoots, because soil temperatures are higher at this season than air temperatures. By the time new leaves are present and demanding water, a new set of absorbing rootlets is ready to meet the demand. Such trees also have the longest possible growing season. Spring plantings (March-April) are sometimes made with considerable success and may be necessary when the acreage to be planted is too large to be completed during the dormant period. Summer plantings are rarely made, although possible. Leaf development tends to precede root development, and the growing season is very short for trees to mature their tissues properly before cold weather.

Methods of Planting

If the tree is set in the grove at the same depth at which it grew in the nursery, the roots are well spread out, the soil is replaced without leaving any air pockets, and sufficient water is supplied to wet thoroughly the whole soil volume occupied by the roots, citrus trees respond well to many methods of planting. The hole may be dug with hoe, shovel, or machine; the tree may be set in place with or without use of a planting board; and water may be applied after each of several increments of backfilling or simultaneously with the soil in a "mudding-in" process. Cost of planting will vary a little from one method to another, but low cost is not economical if the trees do not grow off well.

One method which has proven very satisfactory in commercial practice in the sandy types of soil chiefly found in Florida makes use of the marking stake (or a shortened hoe handle replacing the stake) to make a hole for the taproot, thus assuring that the tree is located accurately. Around this stake a basin is dug to a depth of 6 inches and a radius of a foot. The new tree is set in place with the taproot vertical in the stake hole and the fibrous roots well spread in the basin. A few shovels of soil are backfilled and soaked with water, then a few more shovels of soil are spread, and more water, and finally the hole is completely filled and a water-retaining ring formed above the ground level. The depth of planting is checked and the last buckets of water poured on. It is considered desirable to apply a total of around 10 gallons per tree at planting. In making the backfill, it is desirable to use surface soil for filling in around the roots in the hole and to use the subsoil (if any was removed) for making the retaining ring. It does not seem worthwhile to mix dolomite or colloidal phosphate with the soil used for the backfill. This method is slow, but speed should never substitute for care in planting trees.

A method used on a large commercial scale greatly decreases hand labor in planting. The hole is dug by an auger operated as a tractor attachment, and the spoil dirt is formed at this time into a water ring by machine. Then the nursery tree is "jetted in." A stream of water under considerable pressure from a spray tank or from an auxiliary pump is directed at the location for the taproot by one man, while another places the tree in the proper position. Then the soil is filled in around the roots by washing it down from the sides by the jet of water, assuring freedom from air pockets. This operation is fast and economical.

No fertilizer need be applied at planting, although some growers apply it. There are few or no rootlets able to absorb fertilizer nutrients at this time and if the fertilizer is not very carefully distributed in the hole, tender new rootlets may be injured by coming in contact with a high concentration of salts. Experience has shown that applying fertilizer after removing the banks in spring gives very satisfactory results.

Banking Newly Planted Trees

When winter plantings are made, as in common practice, it is necessary to protect the new trees from possible injury by cold—since they are more susceptible to such injury than they will ever be again—by banking soil around them to a height of a foot or more, preferably as high as the scaffold limbs arise. The important thing is to have the bud union well covered, so that new shoots can develop from the scion even though the top is partly killed back. If the union remains covered by a well-made bank of soil, it will survive any cold known to Florida. It is very desirable to make banks of soil free from decaying vegetation or particles of wood, as the former materials attract ants and the latter attract termites, and both insects may attack the tree under the bank if attracted thus to it. An insecticide may be added to the bank to kill such insects if mechanical means are being used. Otherwise, be sure that any serious infestation is eliminated prior to banking. Wind and rain may carry away some of the banked soil, even to the point of exposing the union, so periodic inspection and replenishment of the banks is necessary. Normally the banks remain in place until the last danger from frost is past in spring, but in an extremely dry winter wilting leaves may indicate the necessity for applying water to the roots, and the banks must be torn down to permit this. Of course, they are at once replaced.

Should low temperature kill the exposed top of the tree, and there ensue a period of unseasonably warm weather, it is necessary to remove the banks because new shoots may start growth on the uninjured stem under the bank. If still covered under these conditions, the bark may become infected by soil fungi, become spongy, and slough off, with eventual death of the tree. A tree unbanked for this reason, with tender new growth starting, is exceptionally sensitive to cold, and the trees *must* be rebanked when further cold is forecast.

In the southern coastal area it is not customary to bank citrus trees for cold protection, as freezing injury occurs so infrequently as to make the practice uneconomical. And in Dade County the nature of the soil precludes the possibility of banking.

6. Bringing Citrus Trees into Production

DURING the first two or three years after planting a citrus tree, the objective of the grower is not to obtain the earliest possible production of fruit but to develop a sturdy tree to good size, so that it may have a long, productive bearing-life later. Little or no formal training is given citrus trees, other than topping them at planting to assure development of low heads. They naturally form strong crotches, the wood is tough, and a sturdy framework of scaffold branches is easily obtained. The grower needs only to aid the growth of the trees by supplying favorable conditions for development. With no crop to consider, all attention can be devoted to promoting vegetative growth. Sometimes growers tend to give minimum attention to these trees because they are not yet returning any income, but this is a mistake which will be regretted a long time because of its adverse effect on the future bearing trees.

By established custom in Florida, citrus trees are classed as nonbearing during the first four years after being planted as yearling trees. Actually they may bear a few fruit during the third or fourth year, but all efforts are directed toward tree growth and any fruit production is incidental. The cultural program for these nonbearing years is quite different from that given to bearing trees. In the following discussion it is assumed that trees have been planted, as is usual, in the dormant season as 1-year-old nursery budlings. Slight adjustments needed to fit this program to trees set out in spring or summer will be obvious and easily made.

Unbanking

When danger of frost is over (February 15 in the southern, March 15 in the northern area), the young trees should have the banks torn down. This is easily done with a large hoe or tractor attachment, removing the soil carefully so as not to injure the bark of the tree and spreading the soil around the tree to make a basin about a yard in diameter for holding water. The last remnants of soil adhering to the tree should be removed by hand to avoid scarring the trunk. Then the basin should be smoothed by hand and any exposed roots covered with soil.

Watering

Watering young trees regularly during the spring dry season is far more important than fertilizing them, as many growers have learned the hard way. Trees lacking adequate mineral nutrients may be stunted in growth temporarily, but young trees suffering from insufficient water are likely to be permanently dead.

Immediately after unbanking the trees and forming the water basins, each tree should be given about 10 gallons of water. The purpose of the basin is to be sure that all this water goes down into the soil around the tree roots, and water must be poured into the basin slowly enough that it does not overflow the sides and run off into the middles. During the remainder of the spring and until summer rains begin in earnest, water should be supplied so as to keep the soil in the root zone from drying out. As a rule of thumb, if two weeks pass without at least an inch of rainfall, trees should again be given 8 to 10 gallons each, with more frequent applications during protracted dry periods. The basin is retained as long as watering is needed. The grower who has a large acreage of young trees must be prepared to water all of them within a 2-week period, so that he is ready to water the first ones again a fortnight after the last watering. In some seasons the spring rainfall may make no watering at all necessary, but the grower must be prepared to deal with the worst situation which may arise, so far as drought is concerned.

Watering in the autumn is seldom needed, for tree roots should have penetrated the soil for some distance, so that they have access to a much larger reserve of water than they had in spring, and constantly decreasing temperatures reduce the rate of water loss from the tree. Except in case of extreme drought, it is better to encourage early winter dormancy by allowing the soil moisture content to become a little low in the fall. Even though summer is the rainy season there are sometimes periods of two or three weeks in summer without rain. Young trees in their first season of growth are not yet provided with root systems able to carry them through such dry periods, and need watering during them almost as much as in spring.

It is in the first year of grove life that watering is chiefly needed; by the second growing season the young tree is more nearly able to take care of its needs for water, but it is still not self-sufficient. All through the four years of its nonbearing stage—indeed, even after it comes into full bearing, too—regular inspection should be made during periods of prolonged drought to see if leaves are wilting and water is needed. When it is evident that watering is necessary, enough water should be applied to soak the soil thoroughly—10, 12, or 15 gallons per tree in successive years—and not just a bucketful to wet the surface soil. Such an application will be lost in a day or two by evaporation and the tree will hardly have the benefit of it longer than that. Such economizing is poor economy.

Fertilization

Citrus trees differ little from any other evergreen trees or shrubs in their need for mineral nutrients, and the fertilizer mixtures generally used for garden ornamentals give satisfactory results with young citrus trees. Such a general garden analysis as 6–6–6 may be used, or the mixtures analyzing 4–7–3 or 4–9–3, with or without minor elements, which are commonly sold as fertilizers for young citrus trees. High analysis mixtures can be used with low poundage per tree and proper care to assure distribution. Small amounts applied at rather frequent intervals have been found more effective than the same total amount applied at intervals of several months. Fertilizer should be spread evenly over

the ground in a circle whose diameter is twice that of the spread of the tree top. For nonbearing trees, especially in the first year or two, the fertilizer is usually spread by hand, although some operators with large acreages of young trees use machine distribution combined with mechanical hoeing. Uneven distribution of fertilizer may result in a relatively high concentration of soluble salts in some places, causing injury to roots under those locations. Large applications are wasteful and may cause the same type of injury even more widely.

A satisfactory schedule of applications during the first three years is to apply fertilizer five times yearly. The first application is made as soon as the banks have been removed and the trees watered in the spring—February 15 in south central Florida. The other four applications follow at intervals of about six weeks—April 1, May 15, July 1, August 15. In the northern part of the citrus area the first application will be a few weeks later, and the succeeding applications will be at intervals of about five weeks. It is considered undesirable to apply fertilizer after September 1 in areas north of the Orlando-Tampa line or after October 1 south of this, lest the young trees be encouraged thereby to continue vegetative growth too long, resulting in poor maturity of tissues at the time the first autumn frost may occur.

The quantity of fertilizer to apply at each application will increase from year to year. In the following schedules, it is assumed that fertilizer with the 4% nitrogen content (such as 4–7–3) commonly employed for young citrus trees is being used. It is equally satisfactory to use fertilizer with 6% or even 8% of nitrogen. For a 6% mixture use two-thirds the number of pounds specified below, and for an 8% mixture use one-half the stated amount. The tree will receive the same amount of nitrogen in all cases and will not know the difference.

SCHEDULE OF FERTILIZER APPLICATIONS FOR YOUNG TREES,
BASED ON A MIXTURE WITH 4% NITROGEN

Year	Feb. 15	April 1	May 15	July 1	August 15
1st	½ lb.	⅝ lb.	¾ lb.	1 lb.	1½ lb.
2nd	1½ lb.	1¾ lb.	2 lb.	2½ lb.	3 lb.
3rd	3 lb.	3½ lb.	4 lb.	4½ lb.	5 lb.

As the quantity per application is increased, the area covered by the fertilizer should increase proportionally, so that the amount of fertilizer for each square foot of soil surface remains fairly constant. Thus with twice the quantity applied in August the second year, the diameter of the circles of application should be half as large again the second year, and in the third year twice as large as the first year.

Trees planted in spring can follow this schedule the first year, except for making each application six weeks later than indicated and omitting the one scheduled for August. The second year they can be fertilized exactly according to schedule. Trees planted in summer will receive only the first two applications, made on July 1 and August 15. In the spring of the next year, they will start with 1 lb. and gradually be given larger amounts until by the end of summer they are receiving almost as much as the dormant-planted trees.

It is not worthwhile trying to adjust the amount applied to each tree to take account of small differences observed in the size of the trees of a single variety. Uniformity in size is desirable and will most nearly be attained by applying to all what is considered satisfactory for the average tree. However, some varieties grow more rapidly than others, and on some soils a given variety grows faster than on others. Such differences between varieties and locations may well warrant a small variation in the amounts applied, increasing them when average growth of a particular block is above normal, decreasing them for markedly slow growth.

The fourth year is a season of transition from the vegetative to the fruiting condition. In February of this year, if good average growth has been made by the grove, apply 5 lbs. of the same 4% mixture, or equivalent amount of higher analysis mixture, and repeat in April and June. Then after the September flush of growth has matured, make a final application in late October or early November of the same amount. Growers often wonder whether to begin in this year the distribution of fertilizer by mechanical spreaders, which are commonly used in fertilizing bearing groves of uniform tree size. As a rough guide, it may be considered that this type of distribution will prove satisfactory when the radius of the crown—the distance from tree trunk to the circle formed by the outermost drip of rain from the canopy—is at least one-fourth the width of the work middle in which the spreader operates.

After this fourth year the trees are considered to be of bearing age, and will be handled as described in the next chapter.

Spraying

The extensive spraying programs needed in bearing groves are not required during the first four years, since little or no fruit is produced and these programs are chiefly concerned with keeping fruit free of pests. However, purple scale, Florida red scale, purple mite, and Texas citrus mite may attack the young trees and cause considerable injury by their feeding if not controlled. Serious defoliation may take place from such infestations, resulting in shortage of food for plant growth and tissue maturity, since leaves are the organs of the plant chiefly concerned with food making. Regular inspection of the trees should be made several times a year, so that infestations may be detected before the pests have built up to serious proportions. Since these are also serious pests of bearing groves and are discussed in more detail in Chapter 8, the reader is referred to that discussion for methods of control.

Fungus diseases are not much of a problem in young groves except for citrus scab. This causes distortion and malformation of foliage and tender shoots (as well as of fruit) of grapefruit, Temples, satsumas, and lemons, but does not attack oranges or tangerines. While rarely as injurious to the trees as the scales and mites, scab does cut down on the machinery for food making and therefore retards growth. Again, control methods are given in Chapter 8.

Minor fertilizer elements, especially zinc and copper, are often not available in the soil in sufficient quantity to provide the young trees adequately and are much more readily available through the leaves, applied as a spray, than through soil applications. Certainly these

FIG. 6. Five-year old 'Valencia' oranges on 'Rough' lemon pictured in 1960. These trees have made excellent growth.
Courtesy Grand Island Nurseries

spray applications should be made whenever deficiency symptoms appear, but in many areas it is well known that young groves are likely to suffer from these nutrient deficiencies. It is much better never to let deficiency symptoms appear than to correct them after they have caused some decrease in growth, and many growers make an annual application of minor elements (preferably combined with control of some pest) each spring. Details of minor element sprays are given in Chapter 8.

Pruning

Undesired sprouts frequently appear on the trunks of young citrus trees and these should be checked several times during the growing season so that they may be removed before they have become large enough to offer much competition to the desired shoots. Various types of trunk wraps have been used to reduce the tendency toward sprout formation. While the food sprouts make may have some value to the trunk and roots, the nutrients and water which they divert from the permanent branches more than compensate for any benefit, and the earlier they are removed the better. Sprouts develop chiefly either from the stock below the union or from the first 6 to 8 inches of the scion above the union, and any sprout arising less than 12 inches from the ground should be removed. While they are still only a few inches long they are easily pushed off with the thumb. If they are allowed to grow larger and form woody tissue, pruning shears will be needed. During the first year the young grove should be gone over every two months from March to November for sprout removal. In succeeding years the interval between inspections may be lengthened, until in the fourth year a single inspection in summer will suffice.

Pruning of the permanent branches is not usually required during the first three years, for the citrus grower is free from the necessity of carefully training the young tree which the grower of apples or peaches feels. In the fourth year, however, just prior to the bearing period, it is well to examine the trees for weak limbs, branches which are parallel to and just above stronger branches, and limbs which lie across and rub against others. Such branches should be pruned out to strengthen the bearing framework. The objective is a well-shaped head with evenly spaced scaffold limbs and space for branches to develop properly without excessive competition from other branches for light. Only a light thinning should be made and not a heavy pruning-out of branches, for while it is always easy to cut out, it is very difficult to persuade the tree to replace branches unadvisedly removed. This shaping and thinning operation should be entrusted to an experienced man and not turned over to a novice as a means of gaining experience.

Painting pruning wounds is usually quite unnecessary in the young grove because of the smaller diameter of branches cut and the rapid healing of such small wounds. However, it is just as important with small limbs as with large ones to make pruning cuts flush with the surface of the parent limb or trunk and not to leave any projecting stub which will heal over only slowly.

Cultivation

Competition from weeds is a much more serious problem with young trees than with old ones because of shallow rooting and small volume of absorbing area. Weed control is important, therefore, in the young grove for rapid and vigorous growth of trees. Weeds are usually controlled by hoeing and this should be done as often as necessary to prevent competition. All weeds should be hoed out in a circle extending from the trunk out at least a foot beyond the spread of the branches.

Much hand labor can be avoided by harrowing the grove along the tree rows in one direction. Cross-harrowing would still further eliminate hoeing—although there is always some hand labor needed close to the tree trunk—but would cut down on the use of the working middles for growing cover crops. A 5- or 6-foot harrow run as close to the trunks as possible, without danger of cutting them, on both sides of the row leaves a wide area in the middles for cover crops. If weeds are small, which is desirable, an acme harrow suffices; for larger weeds a disk harrow set for shallow cutting will be needed. Cultivation, like hoeing, should be done often enough so that weeds never get big, but it has no benefits other than weed control.

Chemical weed control has been the subject of much research during the past decade. A number of herbicides, registered by the Pesticide Regulation Division of the Environmental Protection Agency (EPA), are now available for use in the young, non-bearing grove (and in other situations as discussed in Chapter 10). These can be used along the tree rows to reduce to a minimum the necessity for hand labor and mechanical cultivation. Their proper use reduces root injuries and incidences of foot rot; improves the health and growth of the tree; and reduces slightly the need for water during periods of drought. The single-grain structure of Florida soils does not require cultivation to maintain root development. No specific recommendations are suggested here due to the rapid development of materials and techniques. The grower should seek the advice of extension personnel or industry representatives with respect to available materials and their proper usage.

Instead of the natural growth of weeds in the middles the grower may wish to plant a leguminous cover crop such as hairy indigo or beggarweed, which will produce more beneficial organic matter in the soil later. In September this cover crop—whether natural or planted—should be disked down. If the growth is so heavy that a plow must be used, disk after plowing to level the ground and mix the cover crop better with the soil; but plowing is likely to cut many citrus roots and should be avoided if possible. A standing cover crop is both a fire hazard and a cause of increased cold injury to young trees by preventing free air drainage. On the other hand, wind erosion is much more serious in hilly ground when the soil is not held together by a mass of plant roots. For this reason the middles should be cultivated in autumn only enough to assure freedom from risks of fire or frost, leaving the cover crop roots at a depth of a couple of inches undisturbed.

Ant Control

Several kinds of ants make their nests in grove soils and feed upon the leaves or even the bark of young citrus trees. These pests may injure trees so seriously that they die, and in any case the trees are weakened and delayed in development. In making inspections for other pests, a watch should be kept for lines of ants moving up and down the trees. No attempt need be made (and it would be ineffective if it were attempted) to eradicate the nests of the ants. Satisfactory control can be obtained by placing close around the trunk of the tree a circle of pesticide in powdered form so that it touches both soil and tree. Approved materials for this purpose are now under the regulation of the U.S. Environmental Protection Agency (EPA). Current recommendations can be obtained from extension or industry personnel. This circle may need to be replenished occasionally, as it may be washed away at some point by water running down the trunk, or dead ants may eventually form a bridge over which new hordes of ants can advance to attack the tree.

After the trees have been banked for the winter, ants may form a passageway along the trunk under the bank. It would be possible to prevent this by mixing the poison dust with the soil used for banking, but that would be needlessly expensive to prevent a small number of infestations. When such cases are detected, the bank should be torn down, the tree circled with poison dust, and the bank replaced a few days later.

Cold Protection

Young citrus trees will require banking for the first four years, at least in the colder areas, or until the trunk reaches a diameter of 2 inches. As trees get larger, the thicker bark and the denser canopy decrease the chance of cold injury to the trunk. The last section of chapter 5 covers this operation in detail.

After the tree has been banked, it should be inspected at monthly intervals to make sure the bank has not been washed partly away by rain or wind, that neither ants nor termites infest the banks, that the trees are not suffering from deficit of soil moisture, and particularly after unusually low temperatures, to assess the extent of any cold-damage. Previous discussion has dealt with correction of any of these situations if they are found.

Beginning with the fifth year, when the trees are considered of bearing age, practices in grove management differ somewhat from those outlined above. The following chapters are devoted to the care of bearing trees.

Notes RESET GROVE
VS
Fertilize Same as Young tree
Do all operations need
TO DO at one time
HERB., Sprouting, Fert.
Difficult to manage
(HAVE SYSTEM)
(make sure reset Gets full Dose Nutritional spray)
spray rest w/ rest of Grove.
(clay saves watering)

7. The Bearing Grove and Efficient Performance

DURING the young, nonbearing years of the life of a citrus grove, the practices described in the last chapter had as their primary objective the production of healthy trees with good structural framework, capable of bearing satisfactory crops of fruit for many years. Now we come to consider the trees of bearing age, for which the primary objective of cultural practices is to obtain production of large crops of fruit of satisfactory quality at commercially competitive costs.

Basic Factors in Production

Every grower of citrus fruits on a commercial basis expects to make a profit and certainly should make one. Some growers devote their entire time to the business of citrus production; some grow citrus as one of two or more business activities, not always the major one; and still others have a very minor, even avocational, interest in grove operation. These last include many people who entrust the operation of a relatively small citrus acreage to a production organization, although the owner may spend many pleasant hours in observing his grove and in following the program carried out in it. All these growers have a real financial interest in the production enterprise and so may confidently be expected to have an interest in obtaining sound knowledge of the basic principles involved.

In an earlier part of this book, those factors were set forth which, taken together, define any particular citrus grove, namely, climate, soil, stocks, and scions. No citrus enterprise can be profitable if these factors are not suitable; they are also permanent factors, in that they usually remain the same throughout the life of the grove. Sometimes a grower can vary soil or scion factors, although at some cost, so as to make them more satisfactory for profitable production. Besides these initial and permanent factors limiting possible grove performance, two other factors enter the situation—the biotic environment and the market demand. The biotic factor includes all of the living organisms which may come into the grove and affect the productive capacity of the trees. The market demand may render production unprofitable.

The sum of all of these factors sets the stage for the production program. In the bearing grove it is the way in which the grower utilizes most effectively the permanent factors of grove potential and overcomes the handicaps of the biotic and market limitations which determines profitable operation. The successful grower is the one who maintains his grove at maximum efficiency. This involves keeping a full complement of fundamentally healthy trees in the grove, fertilizing the trees for vigorous growth and heavy crops, spraying to prevent pests from limiting production or lowering grade of fruit, and helping trees to respond properly to fertilizers and sprays by such cultural practices as cultivation, drainage and irrigation, pruning, and frost protection. These seven programs of the production schedule may not all be necessary in every grove, but each of them must be thoroughly understood by the grower, if only so that he may determine intelligently that he does not need it. The first three operations—maintaining full tree strength, fertilizing, and spraying—are soundly based on scientific research, and the influence of both state and federal research findings can be seen in the way they are carried out in all parts of the citrus belt. The last four operations above listed do not rest upon such a firm research base, yet each of them has received much careful study. It may be well to note here that all citrus production practices presently being used have been influenced not only by professional scientists but also by the many intelligent citrus growers who have developed logical and satisfactory programs through years of trial and error. These thoughtful, farsighted men have had a large share in making the citrus industry of Florida strong and progressive.

The extensive accumulation of scientific knowledge about citrus production is of little value unless the grove operator is familiar with it and knows how to apply it. It is for this reason that many growers with small acreages of citrus trees have preferred to entrust the operation of their groves to a professional grove management service, either a cooperative organization or a private one, since the trained and experienced manager can bring this knowledge to bear on grove operation far better than they can. But the work of this manager will be easier, and a more satisfactory business relationship can be enjoyed, if the owner has some understanding of the problems involved.

Grove Rehabilitation

For maximum efficiency of a production unit, or grove, it is essential that every tree location be occupied by a tree and that every tree be healthy. Only with such a grove can the grower utilize approved fertilizer and spray practices with the maximum benefit. If a grove is planted so as to have 70 trees per acre but has by actual count an average of only 65 trees, because the other 5 have died, then the grove can never have more than 93 per cent of its theoretically possible production. Similar loss of productive efficiency applies equally to any planting density. Of course in cases where trees were too closely spaced originally, a considerable number of trees may be removed without decreasing permanent capacity to produce, but we are assuming that trees were properly spaced at the outset. Prompt replacement of trees which have died means higher average returns to the grower.

In many cases, however, the situation involves not only vacancies in the grove but trees which are unthrifty or declining. If these trees are allowed to remain in the grove for many years, growing steadily weaker and yielding less fruit each year, the possible production figure for that grove is also declining annually although costs of production remain the same. In short, possible profits decline slowly but steadily. It is not sufficient, either, simply to remove and replace such trees, once it is clear that they are declining. Unless the reason for the decline can be found and the condition corrected, the replacement may suffer the same fate in its turn. Extensive treatment may be needed before a new tree can safely be set, or a change of stock may be necessary to fit the soil conditions. Furthermore, the tree may be healthy but producing small crops, as often happens on grapefruit stock, or it may be producing large crops of undesirable fruit, as when an unfavorable mutation has been propagated as a strain.

It may be possible to rehabilitate the unsatisfactory tree by surgery or medication if it is diseased, or by topworking if the strain is poor. The grower should consider carefully the cost of such remedial work, to decide whether it will be economically profitable. Often he will decide that the restoration will be more expensive in the long run than replacement, and only replacement can be effective if the stock is poorly chosen or the tree is affected by an incurable disease. Since the grove is the unit of production and each unproductive tree in the grove must be restored to productivity, it is better to think in terms of *grove rehabilitation* than of *tree rehabilitation,* meaning by this term the practices necessary to maintain the full complement of trees of full theoretical productive capacity in the grove at all times. This program should be conducted on a regular basis rather than being deferred until serious losses in production are suffered.

Programs of grove rehabilitation are usually much less important in the early life of the grove than in later years, although instances could be cited of groves in which problems have arisen, even in the first or second year, which resulted in production below standard levels. Climatic factors may be more harmful to young trees than to older ones, and unwise choice of site may be reflected in retarded tree growth early in the life of the grove. Topworking has sometimes been resorted to even before the trees have borne their first crop, when the grower has decided that marketing opportunities are greater with some other variety than he had originally planted. Debilitating disease complexes have sometimes made necessary the entire removal of the first trees planted and their replacement with a different combination of stock and scion.

Because a few trees are always likely to be lost, due to various causes, during the first few years, it is a wise precaution at planting to set out a few "spares" (1 to 2 per cent of the total) in some convenient location within the grove for use in replacement as needed. This assures having trees for resetting of the same type and age as the original trees.

Administration of the grove rehabilitation program involves *identification* of what is wrong with a tree, *appraisal* of the extent of interference with production and of the economic feasibility of treatment, and *action* to remedy the situation. Citrus trees may become and remain too fruitless for profit, and may decline in vegetative vigor or even die, from many diverse causes, among them climatic factors such as freezes, droughts, hurricanes, floods, lightning; soil conditions, especially matters of drainage or salt concentrations; nutritional disorders resulting from deficiency or toxicity of mineral elements; pathogenic diseases of various types, such as bacterial, fungal, or viral; insects, mites, and nematodes; other animal pests, such as rabbits, gophers, or other animals; and even injuries caused by grove machinery. Any or all of these may have an unfavorable effect on the tree and result in lowered vitality and production.

In some cases identification is a very simple matter, while in other cases it may prove very difficult. A tree may respond to several, quite different, causes by exhibiting a condition of "decline," of "dieback," or of "gummosis." Such general symptoms are of little or no diagnostic value, since they indicate only that in some serious way the normal functioning of the tree has been disturbed. A grower is likely to consider a rehabilitation problem as systemic, because the whole tree seems to be affected, when in fact the cause is a localized infection, as with foot rot, or the injury is both local and very limited, as when rabbits have girdled a tree. In making correct identification the grower, or even the production manager, often needs the assistance of the trained personnel of research or regulatory agencies. Even then, there are sometimes cases when it is not possible to be sure of the fundamental cause of decline.

Once the cause has been identified, it becomes possible to make an appraisal of current damage and of its effect on the productive capacity of the grove, as well as determine with some degree of accuracy the expectation of future damage to tree and yield. Such appraisals are necessary as a basis for deciding what corrective measures will be economically worthwhile. Furthermore, in the course of appraisal additional information may be brought out, such as the interrelation of certain factors, which may make the problem more serious than it first appeared, or perhaps less so. Appraisal of water damage, for example, must include not only mapping of the trees to record their relative degree of injury, but also investigation of the water-table situation and the possibility of reducing future damage. If the study shows that it will not be possible to change the depth of the water table, then it may be wise to abandon the site. Similarly a site subject to constantly recurring freeze damage may have to be abandoned if no way can be found to alleviate the damage. The appraisal may suggest that a change in the stock and/or scion variety is desirable, to prevent immediate recurrence of the same problem with new plants, especially if the cause is one of the virus diseases.

It will be noted that appraisal requires study of the problem itself and of contributing environmental conditions, and also of both current damages and economic losses and those of future years. In making these studies, too, specialists may be of great help in determining what measures should be taken.

The final step in the rehabilitation program consists of remedial treatment, if feasible, or of removal of declined or dead trees and resetting of new trees after any underlying causes have been corrected. Declined trees should be taken out just as soon as they fail to return costs of production and appraisal shows that they cannot be restored to health.

Tree removal is best done in October, but might be delayed until December 1 in the case of an early variety so that a last small crop could be gathered. After the larger roots near the surface have been severed with an ax, the tree can easily be pulled out by a tractor. The large roots are then dug out, the hole is filled, and the soil allowed to settle. If any corrective measures are needed, they should be taken at this point. Soil fumigation, or fallowing, before replanting should be considered. Several fumigants are on the market; they should be used as directed on the label with proper interval before planting.

Replacements should be made in December and January, preferably with the same variety already in the block unless a change of the whole block is being made to another variety considered more profitable. Groves can be found in the state in which the grower has used no consistent plan in making replacements, having used for resetting whatever varieties were especially in demand in the given year. As a result, one finds replacements of various ages of different types within the same grove, and many of these trees will never be picked because it is not economically feasible to send pickers into such a grove for the small number of boxes there of each of the different varieties. Under such conditions it is hardly possible to operate a logical production program. If the variety already in the block is an undesirable one for some reason, then the grower should take the best means—topworking or replanting—for changing the whole grove to another variety, an operation which certainly can be fitted into the concept of grove rehabilitation.

Replacements made in old groves should be given the same attention as is given in planting a new grove. Sometimes growers tend to leave the replants to fend for themselves, assuming that the operations employed for the old trees will be adequate for the new. However, because of competition with the older, well-established trees, resets need much more attention. The program for young trees, as set forth in the previous chapter, should be followed for replacements during their first four years in the grove. They should be cared for quite independently of the program followed for the older trees, Broadcast fertilizer applications adequate for the established trees may leave the replants in an impoverished condition, and they may even decline for lack of sufficient nutrients unless they are fertilized individually by hand. Watering may also be needed by these reset trees, because of their much shallower root system, although soil moisture may be adequate for the older trees.

There are benefits to be derived by the grower and operator from a well-conducted program of grove rehabilitation, even beyond the fact that thereby the grove is constantly maintained in efficient producing condition. These additional benefits result from the fact that continued, intelligent observations in the grove will allow them to know the grove more intimately and to meet its needs more certainly for all phases of the production program.

Diseases Often Make Grove Rehabilitation Necessary

As was said before, many causes may bring about a decline in tree growth. Previous chapters have discussed the role of cold and soil moisture in citrus tree growth, and the following chapters will take up nutritional requirements and the diseases controlled by the spray program. There are some serious diseases, however, which are not so controlled and which are often the reason for a progressive decline in tree health and productivity. Among the most important of these are root and foot rots, virus infections, and nematode infestations. Although the cause, or even the type, of the malady is unknown, a discussion of young tree decline is presented here.

YOUNG TREE DECLINE

In 1964 the first symptoms of a new and unknown malady appeared in the flatwood and marsh areas. Affecting trees of five years of age, the symptoms of sparse foliage and premature leaf drop, zinc and manganese chlorotic patterns, and small, misshapen fruit suggested the name "young tree decline." Contrary to tree losses of 1 to 2 per cent per year, which have been considered "normal," random losses and clustered losses rapidly took on the magnitude of a serious threat to the industry. At the beginning it seemed to be a problem of 'Rough' lemon stock when budded to sweet oranges, but other stocks with other scion varieties appear not to be immune. Similar symptoms showed up in groves in the well-drained areas; here the name "sand hill decline" came into vogue. Overall, the name "lemon root decline" suggests the high incidence on 'Rough' lemon stock. No definite cause has, as yet, been identified; active research is being vigorously pursued.

"Scion-rooting," whereby a mound of soil is placed around the tree trunk, usually by mechanical means, to eliminate the understock as roots developed from the scion, has been a practice in some groves. However, even this is worthless after the trees have become weakened. Total replacement has often been required. Loath to continue to use 'Rough' lemon stock, growers are left with choices of sour orange, 'Carrizo' citrange, trifoliate-orange, 'Cleopatra' mandarin, Macrophylla

(*Citrus macrophylla*), or other stocks, even though some of these may be susceptible to the malady. As with the 'Milam', a 'Rough' lemon variant which resists the effects of the burrowing nematode, perhaps a variant type may be found. Hopefully the efforts of numerous scientists and managers will suggest the definite cause and its correction; in the meantime the grove manager must stumble in the dark.

FIG. 7. 'Valencia' on 'Rough' lemon, 9 years old, affected by young tree decline, also known as 'Rough' lemon decline.

Courtesy AREC, Lake Alfred

FOOT AND ROOT ROTS

Foot rot was one of the first diseases to attract attention in Florida. As early as 1876 it was observed to attack sweet orange seedling trees. Practically all of them were then grown on rather low ground, and the resistance to this disease shown by sour orange trees under the same conditions was an important factor in the change from sweet seedling trees for grove planting to sweet orange budded on sour stocks. The disease is caused by either of two species of fungi (*Phy-*

tophthora parasitica or *P. citrophthora*) which are present in the soil and attack susceptible kinds of citrus trees under certain conditions. High soil moisture, mostly found in low, fine-textured soils, is the most common of these conditions. Citrus species vary greatly in susceptibility, the order of decreasing resistance being usually reported as trifoliate-orange, sour orange, 'Cleopatra' mandarin, 'Rough' lemon, grapefruit, sweet orange, acid lime, lemon, sweet lime. 'Rangpur' lime is susceptible. The species more resistant than 'Rough' lemon can be used satisfactorily as stocks on moist soils for more susceptible species so far as foot rot is concerned.

Foot rot is a disease of the base of the trunk, or of the crown, when stocks are susceptible, and can develop when the tissues of susceptible species are near enough to the ground for fungus spores easily to be splashed by raindrops onto stem tissues. Budding very low on resistant stock may still permit infection to occur. Trees budded high but set low in the ground may become diseased. Conversely, trees which are budded high on resistant stock, planted in wet soils set at proper height, and given good air circulation are less likely to be infected. Weeds or low-hanging canopies which prevent free circulation of air are conducive to infection.

When foot rot has developed sufficiently to girdle the trunk one-quarter or more of its circumference, decline becomes quite visible, although it is often seen only on one side of the tree if girdling is partial. The foliage becomes progressively more yellow, due to the cutting off of mineral nutrients from the roots, and finally twigs and branches die back extensively.

If inspection shows decline is due to decay of the bark at the base of the trunk, the life of the tree may be prolonged by cutting away all diseased bark back into healthy bark. The exposed wood should be painted with a safe disinfectant (Avenarius carbolineum) and covered with pruning paint. If the crown and main roots are diseased, the soil should be pulled away to expose all diseased members freely to the air. Then the diseased bark is treated as above. In sandy soils it is well to leave the crown exposed indefinitely unless there is danger of injury by freezing. If grove trees are regularly examined once a year for the first signs of foot rot and remedial measures undertaken, decline symptoms can be forestalled. The value of satisfactory aeration cannot be overemphasized.

Root rot caused by the fungus *Clitocybe tabescens* often produces a decline similar to that of foot rot. The roots are attacked first and the infection may spread to the base of the trunk. If small mushrooms appear from the trunk infections identification is sure, but they do not always develop. The fungus usually is present in pieces of oak or hickory roots, and is largely prevented if all such roots are very thoroughly removed from the soil in preparation for planting citrus trees. This is a disease of high, sandy soils, not of low, heavy soils as foot rot is, and has no relation to the stock used. Infected trees should be removed as soon as productive capacity is lost. Clean area of root debris prior to resetting.

Diplodia root rot may also occur in citrus groves and is often the cause of decline of trees in heavy, wet soil after flooding by very heavy rains or excessive irrigation. The bark of the roots may be seriously decayed before the tree suddenly goes into a marked wilting and decline follows rapidly. By this time the trunk has usually been girdled also. A characteristic feature is the blackening of wood beneath diseased bark. Wilting is likely to occur during a drought, because even the badly diseased roots may still supply sufficient water while it is abundant. No cure is known for the disease, but it may largely be prevented by care in irrigation and by improvement in drainage conditions. After diseased trees are removed, it is desirable to fumigate the soil before replanting.

Virus Troubles

Viruses are composed of nucleoprotein and cause derangement of nucleoproteins of the host cells, resulting in malfunctioning of the metabolic processes. As such, they are responsible for many serious ailments of plants and animals. Citrus trees are now known to suffer from the presence of several virus infections. Some cause a very rapid decline while others cause a very slow decline or merely a permanent dwarfing. The principal virus diseases of citrus trees are psorosis, tristeza, exocortis, and xyloporosis.

Psorosis, also called "scaly bark" in California, is not only the most destructive virus, but is also the one first recognized as a citrus disease. It is worldwide in distribution. Since 1896 a disease characterized by severe scaling of the bark of trunks and large branches has been known, and was for many years supposed to be due to some fungus, though none could be shown present. As the scaly patches become larger, the underlying wood and bark turn brown because of gums and resins produced by diseased cells, and eventually the top begins to die back. Usually trees do not show the scaly-bark symptoms until they are mature, and they may decline slowly for many years. This disease, now called psorosis A, was shown by Fawcett in 1934 to be due to a virus. It is most serious on sweet orange and tangerine, but attacks all types of citrus.

Several other forms of psorosis—psorosis B, blind-pocket psorosis, concave gum psorosis—are known, caused by strains of the same virus, but are much less common than psorosis A. Not all kinds of citrus trees show scaling of bark or exudation of gum as symptoms of infection, but all do show a characteristic chlorotic pattern in the partially grown leaves. This pattern is a symmetrical one of light yellow areas, either small flecks or large blotches, which are translucent when the leaves are held up to the light.

When trees are found to be affected with psorosis there is no known cure. The virus is apparently transmitted almost entirely by buds, not by insects or pruning tools, so there is no hazard to other trees from leaving the diseased trees in place. They should be removed and replaced when declining yield makes them unprofitable. This disease is easily prevented by care in selecting budwood.

In 1952, at the request of forward-thinking growers, a citrus budwood registration program was inaugurated to prevent the extreme losses which often were caused by use of psorosis-infected budwood. Now operated as the Citrus Budwood Registration Bureau of the Division of Plant Industry of the Florida Department of Agriculture and Consumer Services, the program has been expanded to include other virus problems as well as those due to propagable nature. This program now assures the industry seed and budwood sources free of many of these disorders and has added vastly to the knowledge concerning the disorders.

Tristeza first came to serious attention in Florida when it wiped out the citrus industry of Brazil almost completely in the 1940's. It had appeared first in Argentine citrus groves between 1925 and 1930, where it rapidly killed trees on sour orange stock, and had spread to Brazil in 1937. At first the cause was unknown, but eventually it was shown to be a virus infection. The quick decline disease of California was soon found to be due to the same virus. In 1951 the virus was found to be present in Florida also, but in a very mild form so that it has not caused major damage. Nevertheless, in various areas of the state tristeza has been found in apparently more virulent form—to the extent that it has become a source of concern. It is transmitted by budding and grafting and by the three prominent aphids of citrus in Florida (the green citrus aphid, the cotton aphid, and the black citrus aphid). *Toxoptera citricidus,* the brown citrus aphid, which vectored the disease in Brazil, wiping out millions of citrus trees on sour orange stock, does not occur in Florida. The low incidence of tristeza among grapefruit on sour orange rootstock appears to offer some justification for continued use of sour orange stock for grapefruit varieties, but the current popular use of sour orange as a rootstock for sweet oranges and mandarins is discouraged.

This virus causes trouble only when certain stock/scion combinations are involved. Sweet orange and other species used as scions are able to carry virus infections without any injury or evidence of infection unless they are budded on sour orange or other susceptible rootstock. Susceptible rootstocks, other than sour orange, include *Citrus macrophylla*, sweet lime, citremon, grapefruit, some tangelos, and shaddocks. Citranges and trifoliate-orange have been reported as susceptible in some areas. The virus causes a barrier to be set up where the stock and scion tissues join, preventing movement of food from the top to the roots. Before long the stock dies followed by death of the scion. There is no easy way for the grower to identify decline due to tristeza in the field; indicator plants are used in the budwood registration program. However, since rootstocks, except sour orange, used in the commercial citrus belt are generally not susceptible to symptom portrayal, decline limited to sour orange stock may be suspected of being tristeza if no other cause is obvious. The decline progresses through yellowing and various chlorotic patterns of foliage to heavy shedding of leaves and die back of twigs. Usually a piece of the inner bark of sour orange stock, taken just below the bud union, will show a "honeycomb" pattern of fine holes, readily visible under a hand lens.

No practical cure is known for tristeza, and affected trees should be taken out as soon as identification is positive, because the virus may be carried by aphids to other trees. Use of tolerant stocks will avoid injury from this virus in new plantings.

Exocortis affects trifoliate-orange and some of its hybrids (citranges), 'Rangpur' lime, sweet lime, and citron. Like tristeza this virus is present in the twigs of many kinds of citrus fruits all over the state. Injury occurs when it is transmitted to a susceptible stock by use of infected budwood or when exocortis-contaminated pruning tools carry the virus to a susceptible combination. Susceptible and non-susceptible stocks should not be mixed in a commercial grove in order to avoid the latter type of transmission. Exocortis-free budwood should be used regardless of the type of rootstock. The disease has been of minor importance in the Florida citrus belt only because the stocks of susceptible nature were used to such a limited extent. However, several of the newer stocks and many in experimental plantings are exocortis-susceptible.

The symptoms of exocortis are extensive scaling of the bark of the stock below the union and/or a stunting of scion growth. Stunting may be the only symptom and is often extremely severe, but when scaling of bark occurs, stunting is always present also. It is now recognized that reports soon after 1900 of dwarfing of some varieties of sweet orange on trifoliate stock while others made normal growth was due to lack of this virus in budwood of the normal group. No cure for infected trees is known. The disease should be avoided at planting time, with an understanding of the hazards, by proper selection of stocks and scions. It was early thought that satsuma budwood was always free of exocortis in Florida. However, all the old-line satsumas tested in the budwood registration program have given positive exocortis reaction with citron indicators; on trifoliate stock these lines produce dwarfed trees that show no scaling, or almost no scaling. A number of nucellar satsuma selections are now available that are free of exocortis and grow normally on trifoliate stock.

Xyloporosis is a virus disease which apparently is transmitted only by buds. The virus occurs in symptomless form in tolerant combinations of stock and scion, but symptoms develop when xyloporosis-infected budwood is propagated on *Citrus macrophylla,* 'Rangpur' lime, or 'Palestine' sweet lime. Furthermore, 'Orlando' and 'Minneola' tangelos, 'Murcott,' satsuma, 'Robinson', 'Osceola', 'Lee', and probably 'Page' and 'Nova' are susceptible, as scion varieties, regardless of the rootstock type used. The characteristic symptom of infection, found on the trunk of sweet lime stock or tangelo scion, is the development of small pits in the outer surface of the wood into which little projections from the inner side of the bark fit. This is easily seen when a flap of bark is cut and lifted just below the bud union (sweet lime) or just above it (tangelo).

Decline occurs in infected trees. Symptoms are bark scaling, gum infiltration, diminution in leaf size, chlorotic patterns of mineral deficiency, leaf fall, and stunting, often with flattened tops, small fruit, and poor yields. Infected trees are an economic loss in a few years and must be replaced. To avoid this disaster, xyloporosis-free budwood should be used (which must also carry freedom from psorosis as a nursery requirement); it is now available for every citrus variety propagated in Florida. Sweet lime, once eliminated in Florida because of this virus, along with other susceptible stocks, can be used with this budwood. The desire of the industry will continue to be the total elimination of xyloporosis.

Nematode Infestations

For many years Florida citrus growers, like other horticulturists, thought only of the species causing root knot (to which citrus trees are immune) when nematodes were mentioned. This situation was radically changed in 1946 when the citrus nematode was reported widely spread in Florida groves and was strongly suspected of causing spreading decline, a mysterious disease which had baffled research men for years. In 1953, however, it was definitely shown to be another species, the burrowing nematode, which was the cause of spreading decline. As the name implies, spreading decline is characterized by a condition of decline which slowly but steadily spreads through a grove, at an average rate of 50 feet a year from the point or points of its first appearance. The symptoms shown by the aboveground portion of the tree are very general and common to decline from many other causes —steady decrease in vigor, size of fruit, yield, and amount of foliage— but the infestation of the roots by the nematode and the progressive spread from tree to adjacent tree are distinctive. The disease readily spreads across roads up to 150 feet in width, the nematode moving on tree roots that grow beneath and contact roots from trees on the other side of the road. Only the young, feeder roots are attacked, but eventually the tree may lose up to half of its absorbing system. Nematode infection potential is greatest in the subsoil immediately below the topsoil, 3 to 10 inches below the surface. Therefore, greatest damage can occur at any depth, usually below 10 inches.

The burrowing nematode, *Radopholus similis*, is widely distributed in tropical areas of Asia and the Americas, and attacks a large number of plant species. It is only in Florida that a strain of the nematode is known to attack citrus trees, although citrus is commonly grown in many countries where the nematode is a serious pest on other crops. The nematode moves through the soil in search of new roots when the one it has been feeding on becomes diseased and damaged. As citrus roots extend out into the soil the pest can move from the roots of one tree to those of another. If any of its numerous other hosts, particularly among the weeds, are in the intervening area, they can also become infected and result in further spread. The nematode is widely distributed in the soil throughout the citrus area of Florida but it is only in the deep, well-drained sands, such as on the Ridge, that heaviest damage occurs and the typical symptoms of spreading decline are found.

Control involves preventing introduction to new groves and elimination from infested ones. The first objective is not too difficult if the nematode is not already present in the soil and is not introduced on roots of other hosts or possibly on cultivation equipment. Citrus nursery stock can be obtained which is certified to be free of infestation. Hot water treatment, by immersing the root system for 10 minutes in water kept at 122°F., has been found effective for eliminating nematodes from the roots of young budlings. Commercial nursery sites must be approved, after determination of freedom from nematodes, by the Division of Plant Industry. In home plantings of citrus there is danger of planting ornamental shrubs or herbaceous perennials which may be infested with this parasite near the citrus trees, thus providing a source of infestation for the citrus trees. Ornamentals should also be certified free, or rid of nematodes by the hot-water treatment, before being planted anywhere near citrus trees.

Citrus stocks which show resistance to the nematode have actively been sought with some success. 'Milam', a citrus hybrid of unknown parentage, was discovered in 1954 in a Polk County grove which had become infested with burrowing nematodes. The individual tree from which this variety originated showed no symptoms of spreading decline. It has been used as a rootstock for the past 10 years with a small, but steady, increase in nursery propagations; about 6 per cent of trees are now currently budded on it. It is particularly recommended as a "biological barrier" to prevent spread of the nematode from known locations. Two other discoveries in infested groves have not had widespread acceptance. The 'Estes', a tolerant 'Rough' lemon type, has never become popular, while the 'Ridge Pineapple', though resistant, has, like other sweet seedlings, been decidedly unpopular because of foot rot and drought susceptibilities. Some of the navel oranges, such as the 'Algerian', have shown resistance but are ruled out as rootstocks because of foot rot and drought problems and their generally seedless fruit. 'Carrizo' citrange owes some of its current popularity to the fact that it is tolerant, although not resistant, to the burrowing nematode.

Eliminating the nematode from infested groves is still a somewhat uncertain matter. The recommended procedure is to map the grove and locate all of the trees showing infection after sampling. The nematode may be infesting roots of trees for 2 to 4 rows in each direction from the visible limits of decline, and so all affected trees and marginal trees for 2 or more rows are removed. As many as possible of the roots left in the soil are dug up by deep plowing, and both trees and roots are burned to avoid spreading the parasite further. Then the soil is leveled and fumigated with D-D mixture at the rate of 60 gallons per acre. The fumigant should be injected 12 inches deep to avoid its loss to the air. No replanting should be done for 2 years and no other host plants, including the many weeds found in groves, should be allowed to grow in the treated area, so that any nematodes surviving fumigation will die of starvation. If no indications of infestation appear after two years in the trees nearest the treated area, it is hoped that the nematodes have been eliminated from this location.

Buffer zones have been established to prevent the spread of the burrowing nematode from one area to another in a grove. Buffers are maintained around the known infested areas, treated with fumigants, and kept free of grass and weeds by use of herbicides. This program seeks to limit the advance of the nematode through the grove.

Another approach to control in infested groves is quite promising but still tentative. This involves treatment of the soil with certain fumigants which can be applied around living trees without injury to their roots, yet will kill the nematodes, as is possible with the citrus nematode. If this promise is realized in practice, it will permit cure of diseased trees. These measures, now available, can only slow down the rate of spread of the disease. The permanent solution to the problem, however, can be achieved only with resistant rootstocks. Some species of the African citrus relative *Citropsis* show such resistance as do the varieties mentioned above.

The citrus nematode, *Tylenchulus semipenetrans,* occurs throughout the world wherever citrus has been grown for any length of time. It was considered a minor problem after its discovery in California in 1912 and in Florida about 1913. Its widespread distribution in Florida in 1946 caused much alarm, but because citrus trees can endure quite heavy infestations without showing symptoms, so long as growing conditions are good, it maintained its status as a minor pest. However, there is now good evidence that the citrus nematode is causing considerably more damage than was thought earlier even though the symptoms are not as bizarre and devastating as those of burrowing nematode infestations. All types of stocks have shown damage with the exception of the trifoliate-orange which appears to exhibit a resistance to this pest. Badly infested trees appear undernourished, with small, slightly yellow leaves, small and often unmarketable fruit, and general stunting and deterioration. Contrary to the current practices with burrowing nematodes, citrus nematodes can be reduced while the trees are in place by the use of DBCP (Nemagon, Fumazone) as a soil fumigant. Furthermore, for planting new groves, or resetting trees in old groves, nursery trees free of infestation should be used; soil fumigation of reset areas prior to replanting is advised.

The root-lesion or meadow nematode, *Pratylenchus coffeae,* has caused damage to citrus in a few cases. This pest is very limited in distribution in Florida citrus and should not become a serious problem if nurseries are kept free and resets are treated prior to replanting, as is now required for the burrowing nematode.

8. Fertilizing Bearing Citrus Trees

THE soils on which citrus trees are grown in Florida are low in both native fertility and in ability to retain fertilizer elements applied to them. Consequently the trees can obtain an adequate supply of nutrients for satisfactory growth and production only if they are provided annually with fertilizers in proper amount and kind. Cost of fertilizer and its application constitutes nearly half of the total production costs, and all other cultural practices are in vain if fertilization is not properly done. To a large extent the vigor of the tree and the quantity and quality of the crop are dependent on the fertilizer it receives. Understanding of how to fertilize citrus trees properly is, therefore, of greatest importance to a citrus grower. In passing it may be noted, however, that the soil conditions which make the annual fertilizer bill so large are clouds with silver linings. These very conditions of low capacity to hold nutrient elements and of heavy rainfall in Florida also assure the grower of freedom from worry about accumulation of soluble salts in the grove soils to the toxic level—a condition seriously troubling citrus growers in some parts of the world.

The Essential Nutrient Elements

Fifteen chemical elements have been found essential to satisfactory growth and functioning of citrus trees, as well as of most other green plants. Three of them are adequately provided in any environment suited to tree growth and are largely beyond the control of the grower so far as their nutrient functions are concerned. The other twelve are the fertilizer elements.

The three elements provided by Nature are carbon, hydrogen, and oxygen, which together make up about 95 per cent of the dry weight of a tree. Carbon and oxygen are taken into the leaves as carbon dioxide from the air, and there they combine with hydrogen, taken in as water by the roots, to produce carbohydrates. This process, called photosynthesis, can take place only in green cells (containing chlorophyll) in the presence of light, which supplies the energy stored up in the carbohydrates. These, together with proteins, fats, and other organic compounds derived from them, are the true foods of the plant, used to make new tissues and provide energy for growth.

The twelve fertilizer elements, often miscalled "plant foods," may be present in the soil in adequate amounts, but under Florida conditions are almost always present in inadequate supply and must be provided by the grower. Some of them are utilized by the tree in making such plant foods as proteins while others seem to have some regulatory function. When any of them is not present in sufficient amount, the tree is restricted in its functioning. In case of severe shortage of an element there is usually a characteristic deficiency symptom exhibited by the leaves (or other organs) and this symptom usually persists until the deficiency is corrected. Often, however, two or three elements are deficient in varying degrees simultaneously, and the resulting deficiency symptoms are not so easily recognizable. Conversely, excessive amounts of some elements may be present in the soil and may also prevent the tree from functioning properly. It is important that the fertilizer elements be present in proper balance as well as in certain minimal quantities.

These fertilizer elements are divided into two groups, often termed *major* and *minor* but probably better known as *macronutrient* and *micronutrient* elements. The terms *major* and *minor* convey the implication that one group is of greater importance than the other, whereas both groups are equally essential for plant growth and fruit production, and with both groups the amount needed of any element is affected by the amount of the other elements which are present. The macronutrient elements are those needed by the plant in rather large amounts, while the micronutrient elements are needed only in very small amounts. These latter are sometimes termed *trace* elements, a term to which no objection can be taken but for which there is no convenient contrasting term for the other group. Occasionally the micronutrient group is referred to as the *rare elements*, but this term has a very different meaning to the chemist.

The macronutrient elements are in turn usually divided into two groups, the *primary* elements (nitrogen, phosphorus, and potassium) and the *secondary* elements (calcium, magnesium, and sulfur). The distinction between primary and secondary has historic roots, reflecting the opinion of plant scientists of the last century that the elements of first and great importance for plant growth were nitrogen, phosphorus, and potash. These were the elements which produced large increases in plant yield, whereas the other three elements seemed to exert some sort of subordinate effect and so were called secondary. We recognize now that these responses merely reflected the degree of deficiency commonly encountered. Today the terms have no significance as to the relative importance of these six elements, although for most crops the fertilizer mixture is likely to contain intended amounts of only the primary elements. In Florida, magnesium has become as "primary" as the big three, but calcium and sulfur are hardly given any thought in nutrient planning, although important in connection with control of soil reaction.

The micronutrient elements of interest to the citrus grower are iron, copper, manganese, zinc, boron, and molybdenum. In general they are needed by plants only in amounts about one-hundredth or less of the amounts needed of the macronutrient elements. Until about 1920 none of these elements except iron was known to be needed by plants, because it was so difficult to purify sources of the macronutrients sufficiently that they did not contain these micronutrient elements as impurities in adequate amounts for the very slight requirements of plants. Today all of the above micronutrient elements must sometimes be applied to citrus trees in Florida for satisfactory growth and fruiting. That they are not always needed as fertilizer components is due to their presence in some soils in sufficiently available supply and to their presence in organic matter added to the soil or in certain sources used to supply macronutrient elements.

In general the micronutrient elements are needed by the plant to form organic catalysts, or enzymes, in combination with a protein. Only minute amounts of them are needed because enzymes in very small amounts control chemical reactions within plant cells between large amounts of plant foods or other compounds and are not themselves used up in the process. Whereas the macronutrient elements (except potassium) enter into the composition of living plant substance, cell walls, and stored food, the micronutrients do not become part of these products, but are used over and over again in their formation or degradation.

Deficiencies of micronutrients usually result in rather distinctive patterns of leaf chlorosis, the intensity of the chlorosis being proportional to the severity of the deficiency. Sometimes twigs and fruits may also exhibit characteristic symptoms of deficiency. On the other hand, deficiencies of the macronutrient elements do not usually produce characteristic symptoms in leaves, twigs, or fruits of comparable distinctiveness unless the deficiency is quite acute. Excessive amounts of the macronutrient elements are often merely tolerated by the plant or result in lowered quality of fruit, whereas excess of the micronutrient elements often results in evident injury or toxicity, and the amounts producing toxic symptoms are usually still small as compared to the quantities of macronutrients required.

As a final important contrast, it should be noted that there is usually a relationship between the amount of the macronutrient elements available (up to an optimum) and the amount of tree growth and fruit production. For citrus trees the levels of nitrogen and magnesium availability are especially related to yields. There is no such relationship with the micronutrient elements. A certain level is necessary to avoid deficiency symptoms but amounts in excess of this have no effect on tree performance until they reach toxic levels.

The Nutrient Elements in Fertilizer Practice

Nitrogen—This may be considered as the "balance wheel" of citrus production for it is the element whose level of availability will, in general, determine the yield throughout the normal range of tree response. This is because nitrogen is used in large amounts for production of new plant cells and tissues, and when it is in short supply all new growth is limited. When nitrogen is supplied in slightly subnormal amounts over a long period of time (chronic mild deficiency), the tree responds by reduced growth and fruit production not accompanied by any specific foliar symptoms of the deficiency. If the nitrogen supply is cut off or greatly reduced for trees which have previously enjoyed an adequate supply, the tree responds rather soon (depending on residual supplies of organic nitrogen) with a marked chlorosis. The older leaves become yellowed and drop early, and later the new leaves also are unable to develop a good green color, followed in cases of persistent severe shortage of nitrogen by defoliation, dropping of fruit, and death of shoots. Above levels of visible deficiency, shoot growth and yield increase steadily with nitrogen supply up to a maximum value, unless some other element is so limiting that the tree cannot utilize the increased nitrogen efficiently. Since nitrogen level is more closely correlated with growth and fruiting than any other element, it usually determines the need for other elements to a large extent. In fertilizer formulas or ratios, nitrogen is always listed first.

The amount of nitrogen needed for a bearing citrus tree is most satisfactorily based on the potential yield. While this amount varies somewhat with differences in the type of soil on which the trees are growing, in general it averages about 0.4 pound of nitrogen per box of fruit for oranges and 0.3 pound for grapefruit.

A high level of available nitrogen increases vegetative growth, flowering, fruit yield, and percentage of acidity in the juice; it causes a slight decrease in percentage of soluble solids in the juice but greatly increases the total amount of solids in the crop because of the greater yields. Level of nitrogen has no effect on rind thickness, although a high level may delay coloring of the rind somewhat. A high nitrogen level also decreases intake by the tree of phosphorus, potassium, calcium, copper, and boron, making it necessary to provide higher levels of these elements, but it increases the intake of magnesium, iron, and manganese. Excess levels of nitrogen are of no value and should be avoided.

It used to be considered that the most important items in choosing a source of nitrogen were the rates of availability and leaching, but it is now evident that there are other factors involved of perhaps greater significance. Natural organic forms seem less satisfactory, even apart from their much greater cost, than inorganic forms, because of lower quantity and quality of crop. Only in case no micronutrient elements are supplied deliberately in fertilizing would organic sources have any advantage. The availability of organic forms of nitrogen over a long period of time, as compared with the readily leached inorganic forms, is often cited as an advantage, but research has shown that this has little practical value, perhaps because of the unexpected ability of citrus trees to take in and store up nitrogen during rather short periods of absorption.

Choice of the two principal inorganic forms—nitrate and ammoniacal—seems closely related to soil reaction. When this can be maintained fairly well within the range of pH 6.0 to 7.0, the ammonia form seems to produce slightly higher yield and quality. At pH values of 5.0 or below, the nitrate form is superior. However, exclusive use of ammoniacal forms requires more frequent checking of soil reaction and correction before it gets down to pH 5, since trees decline markedly as the reaction becomes more acid. A mixture of nitrate and ammonia forms is a very satisfactory compromise in every way; synthetic ammonium nitrate, half nitrate and half ammonium nitrogen, is a regular and low cost source material. The differences in tree response are rather slight, however, if pH is controlled, and price per unit of nitrogen is most often the factor determining which source is used. Nitrogen in either form is readily leached from sandy soil, so that it is not possible to build up a supply in the soil for future use except for rather short periods by use of green manures.

Nitrogen is absorbed by citrus trees in Florida at all times of year. The rate of intake in winter is about half of the intake rate in summer, but as leaching is also much less in winter, nitrogen applications are about as effective at one season of the year as at another.

Phosphorus—This element is analyzed by the chemist and guaranteed on the fertilizer tag as equivalent P_2O_5, usually spoken of as available phosphoric acid, but is applied usually as a superphosphate. In spite of a fairly low natural level in many Florida sandy soils, it has not been possible to demonstrate symptoms of phosphorus deficiency in citrus trees grown without added phosphate except on a muck soil, the Davie series, in the eastern Everglades—a soil type which is not extensively planted to citrus. Some of the soils used for citrus plantings have a considerable native phosphate content, and even the soils low in phosphate fix in the top foot or two nearly all phosphorus applied to them. There would be rather little reason to make applications of phosphorus, at least after the first few years of grove life, if it were not for the benefit derived by cover crops and for the effect of accumulated phosphate residue in preventing rapid loss of applied magnesium. In former years phosphorus was applied very liberally, half of each ton of fertilizer often being superphosphate. The cost of phosphorus was much lower than any other element, the superphosphate enabled cheaper forms of liquid nitrogen sources to be used in preparing fertilizer mixtures by its reaction with them, and no evidences of toxicity from the large residues of phosphate fixed in the soil have ever been found in the general citrus belt of Florida. It is abundantly clear, however, that much of the money spent in supplying these large amounts of phosphorus through many years has been wasted. A high phosphorus level actually decreases the percentages of solids and acidity in juice and delays coloring of Valencias. Approximately 0.05 pound of available phosphoric acid is required per box of fruit. Groves which have been bearing for a dozen years and have been supplied with phosphorus regularly will show no benefit from further applications, and even young groves will really benefit only from a small fraction of the amounts which have been customary. Phosphorus is absorbed about ten times as rapidly by citrus trees during summer and early fall as during winter and early spring.

Potassium—This is calculated and guaranteed on the fertilizer tag as the equivalent K_2O, or potash, although it is never present in this form in fertilizers. Until 1945 it was standard fertilizer practice to include more than twice as much potash as nitrogen in the mixtures applied to citrus trees, such as the widely popular 3-8-8 analysis of that time. The high potash level was believed to produce fruit of better external quality as well as to cause early maturity of stem tissues in autumn. Now it is known that in oranges such high levels of potash actually produce large size, coarse skin texture, and slow loss of green color as fruit matures—conditions long supposed to result from excess of nitrogen. Ratios of nitrogen to potash today are usually 1:1, although there is evidence suggesting potash could well be lower still. Since growers have always used some potash in citrus fertilizers and have almost invariably used excessive amounts, it is not surprising that potassium deficiency is unknown in commercial citrus practice. Experimental work has shown that such deficiencies can be produced by leaving potash out of the fertilizer program for several years, for unlike phosphorus, potassium is at a very low level naturally in all soils of Florida and applied amounts tend to be leached rather rapidly from the soil.

Approximately 0.4 pound of potash per box is recommended annually for oranges and 0.3 pound for grapefruit, just as for nitrogen. All of the usual source materials seem to supply available potash about equally well. There is practically no absorption of potassium by citrus trees during cold weather, but the element is readily stored for future use by trees.

Magnesium—It is only since 1930 that this element has assumed an important place in fertilization of citrus trees in Florida. After the discovery that bronzing of grapefruit leaves—a chlorotic pattern in which green was gradually replaced by a dark yellow color—was due to deficiency of magnesium, it was recognized that deficiency of this element often was a limiting factor to yields and made trees very subject to injury by cold. Today magnesium compounds are regularly supplied to augment the very low natural supplies in the soils of Florida. Dolomitic limestone, a naturally occurring mixture of calcium and magnesium carbonates, serves as well as ordinary calcic limestone to correct acidity and at the same time puts large amounts of slowly available magnesium in the soil. About 0.3 pound of MgO, magnesia, the form in which the element is guaranteed, is needed per box of citrus fruit, and in general about half of this will be provided when dolomite is used as an amendment to control soil reaction. The balance of the needed magnesium, or all of it when calcic limestone is used, is supplied by a readily available form in the fertilizer mixture. With regular use of dolomite to maintain the soil pH between 6.0 and 7.0, magnesium may be eliminated from ground applications of fertilizer. On alkaline soils, foliar applications of magnesium nitrate are effective. Seedy varieties respond to magnesium applications to a greater extent than seedless ones, because seeds store large amounts of magnesium and remove it from the leaves, where it is needed to produce the green coloring pigment, chlorophyll. So long as magnesium is provided in amounts which prevent deficiency symptoms, there is no increase in growth or yield as magnesium levels increase. Absorption of magnesium, like that of potassium, ceases almost completely in winter.

Calcium—While this element is needed in large amounts by green plants, the citrus grower need not be concerned about it for its nutrient role, because the supply is always sure to be adequate. This is true because of the regular need of large amounts of calcium carbonate for controlling soil acidity, as well as the further large amounts of calcium supplied in superphosphate.

Sulfur—As with calcium, deficiency of sulfur has never been found in citrus plantings in Florida and probably never will be, in spite of the rather large amounts needed by trees and the low level of this element naturally in Florida soils. There are two good reasons for this. Large quantities of combined sulfur in available form are regularly supplied in the materials used primarily to provide other elements, especially in superphosphate but also as sulfates of the heavy metals and ammonia. And further large amounts of free sulfur are commonly applied to the foliage and fruit for control of rust mite, eventually being washed down to the ground by rain and undergoing changes in the soil which make it available as a nutrient. The problem of sulfur for the citrus grower is not one of possible deficiency or excess of sulfur as a nutrient element, but of the effect of pesticidal sulfur on soil reaction, as will be discussed later.

Iron—Chlorotic patterns due to iron deficiency have been recognized far longer than any others, and always appear first on the young shoots. A striking feature is that the soil may have a high content of iron, even of available iron, yet plants may either not be able to absorb it or be unable to utilize it after absorption. Iron deficiency occurs often in plants growing in calcareous soils, in waterlogged soils, or in soils very low in organic matter—the so-called "sand-soaks." About

1950, citrus growers were startled to find symptoms of iron deficiency in trees on high sandy soils which had been considered well supplied with available iron. Research showed that in spite of good amounts of iron, plants were unable to utilize it because large amounts of copper had accumulated in the top 6 inches of soil and made iron unavailable. The copper residues, sometimes as much as 750 pounds per acre, had resulted from the combination of copper fungicidal sprays and copper added in fertilizer mixtures. Accumulations of either manganese or zinc also have a depressing effect on iron availability, but neither of these metals has yet reached the level in citrus groves where its effects are as serious as those of copper have been. For some reason, application of iron by spraying has not proven satisfactory on citrus trees, and soil applications merely increase the already large amount of unavailable iron in the soil. The solution to the problem has been the use of chelated forms of iron, which put iron into the soil in an organic combination immune to the effects of copper but available to plant roots. On acid sandy soils any chelated iron, applied at a rate to supply 20 grams of metallic iron per tree, gives prompt recovery. On calcareous soils it is more difficult to overcome the factors making iron unavailable. One or two types of iron chelate (hydroxy-chelates) are often but not always effective, and when they are, larger amounts of metallic iron (50 grams) per tree and a longer time for tree response are needed.

Copper—The conditions known variously as red rust, exanthema, dieback, or ammoniation were noted in Florida citrus trees as early as 1875, and were attributed to various causes, especially to use of ammonia or manure. About 1912 it was discovered that application of copper sulfate ("bluestone") to the ground under citrus trees cured this disease, although no one then thought copper could be essential to plants, but it was not until the late 1930's that regular use of copper eliminated dieback as a serious problem. Growers and research workers alike tended to overlook the fact that when copper sprays were regularly used for fungus control, no other measures were needed to prevent dieback. In planning fertilizer mixtures to avoid mineral deficiencies, copper was included without regard to spray residues, and after some ten or fifteen years of double dosage, the excessive accumulations of copper in grove soil brought about the iron deficiencies above discussed. Today it is recognized that no copper should be included in fertilizers if copper sprays are regularly applied or if the grove soil has over 50 pounds of copper per acre in the top 6 inches. In young groves not yet bearing or in groves where copper sprays are not employed, about one-thirtieth as much copper (as CuO) should be included in the fertilizer mixture as the amount of nitrogen. Soil applications should cease as soon as the accumulated copper in the top half-foot reaches 50 pounds.

Zinc—The effects of zinc deficiency are most often expressed in citrus trees as *frenching* of foliage, a severe chlorosis in which leaf tissue becomes nearly or quite white except for green veins. At the same time leaf size is progressively smaller as the deficiency becomes more severe, and shoot internodes become shorter, giving a rosette effect. Recognition of this condition about 1934 as due to zinc deficiency led to realization that other metallic elements might also be deficient. Since frenching severely decreases the leaf area for food making, zinc-deficient trees make very restricted growth and bear little fruit. Citrus trees are not readily able to obtain adequate supplies of zinc from soil applications, but the metal is readily absorbed from sprays. After visible symptoms of deficiency have been corrected, smaller quantities of zinc are applied in an annual spray to prevent recurrence of the deficiency. If these preventative sprays are used from the start of the grove, there need never be any deficiency.

Manganese—A mild form of chlorosis has since about 1932 been associated with deficiency of manganese on acid, sandy soils. The marl chlorosis or marl frenching found on calcareous soils is the result of combined deficiencies of manganese and zinc, and sometimes of iron also. In the 1930's serious cases of manganese deficiency were found in many groves on acid soil because the pH had been allowed to drop well below 5.0 and manganese had been leached out. With maintenance of the reaction above pH 5.5 manganese accumulates in the soil. Because of the widespread manganese deficiencies, this element was commonly added in fertilizers from 1935 on, but by 1950 excessive accumulations of it were becoming apparent. A temporary mild deficiency pattern on new shoots is not detrimental to the growth or fruiting of citrus trees. Only in case the deficiency symptoms persist throughout the year should corrective measures be taken. On sandy soils manganese sulphate, at 2 pounds per tree, is quickly effective from ground application, but on calcareous soils this is not true. On such soils manganese sulphate may be applied as a spray, but the effect is only noted for one season. More lasting results are obtained by mixing 3 pounds each of manganese sulphate and calcium chloride and applying this mixture in about ten piles of ½ pound each, distributed around the tree under the outer edge of the canopy. Because of the effect of accumulated manganese on iron availability, it is well to be cautious about including manganese in regular fertilizer mixtures.

Boron—Florida growers first became aware of this element because of the toxic effect of boron impurities in certain potash salts. Occasionally, also, boron toxicity occurred in trees where borax-dipped field boxes from packinghouses were stacked around them. Once the source of the injury was identified and further additions stopped, boron could be eliminated from the soil by liming and irrigation, or more slowly by the leaching action of normal rainfall. Deficiency of boron is not often found in Florida, but does occur sometimes when growers use only high analysis fertilizers, or following a prolonged drought, or most commonly when arsenic sprays have been applied. Boron is easily available to citrus trees from applications of a soluble borate either on the ground or as spray, the latter method giving quicker response. There is only a very small difference between amounts of boron which are beneficial and those which are toxic, so that unusual care is needed in applying this element. It should never be applied in the fertilizer mixture and as a spray in the same year, and should not be applied at all unless there is some evidence of need. Trees on 'Rough' lemon show deficiency more slowly than those on sweet orange, and these in turn more slowly than those on sour orange stock. Where one desires to avoid any possibility of deficiency arising, boron may be included in the fertilizer at about one-hundredth the amount of B_2O_3 as of nitrogen.

Molybdenum—Ever since the turn of the century a peculiar chlorosis of citrus leaves was noted, later termed "yellow spot," which did not respond to any of the treatments which corrected other chlorotic symptoms. Only since 1950 has it been realized that yellow spot is a symptom of molybdenum deficiency. This element is less available in acid than in slightly alkaline soils, unlike all other heavy metals, and since deficiency occurs usually on soils which have been allowed to become undesirably acid, soil applications of molybdenum compounds do not relieve the deficiency. However, molybdenum sprays, using 10 ounces of sodium molybdate per 500 gallons in severe cases and 5 ounces for milder deficiencies, give prompt correction. Molybdenum sprays should be used only when the yellow spot condition is present, for the element is toxic to cattle in low concentration and much citrus peel goes into cattle feeds. Groves in which the soil pH is maintained above 6.0 are not likely to exhibit molybdenum deficiency.

Soil Reaction and Nutrient Availability

The term *soil reaction* refers to the acidity or alkalinity of the soil and is expressed in units of pH. In Florida these values range from pH 3.0 (an exceedingly acid condition) to about pH 8.5 (a moderately alkaline condition). The neutral point on the scale, where neither acidity nor alkalinity exists, is pH 7, and each scale unit represents a tenfold change. Thus pH 5.0 is ten times as acid as pH 6.0. Two general situations are found in the state as regards soil reaction. Groves on alkaline soils are found in some sections, almost all in coastal areas, but the great majority of groves are planted on acid sands.

The alkaline soils of Florida are all calcareous, that is, composed largely of calcium carbonate. They have been leached of all bases other than calcium, and the very high content of this element in the soil solution makes magnesium and potassium added in fertilizers less available, so that they must be supplied in larger amounts than suffice on acid soils. All the micronutrient elements except molybdenum are only very slightly available at the high pH values of these calcareous soils and are usually not made available by soil applications but must be applied as sprays for the trees to benefit. As noted previously, molybdenum is more available in alkaline than in acid soils, and iron is not satisfactorily absorbed by citrus trees as sprays. Even chelates do not always make iron satisfactorily available on these soils. Sulfur cannot well be used to change the reaction of calcareous soils, as it can be in acid soils, because its only effect would be to cause decomposition of calcium carbonate to form soluble calcium salts, and this would simply cause the soil to disappear when it consisted almost wholly of calcium carbonate, as marl does.

Some citrus groves are on soils in which the topsoil is acid but the subsoil is calcareous. Most of the absorbing roots are in the topsoil and such groves usually need the same treatment as groves without alkaline subsoil, except that the treatment will be modified to take account of the depth of the acid layer.

The well-drained, sandy soils are naturally acid—sometimes too acid for citrus trees to grow on them—and they tend to increase in acidity under grove culture. Partly this is due to the loss of basic elements of fertilizers from the soil through absorption by roots and through leaching by heavy rainfall, but the greatest cause of increasing acidity is the accumulation of sulfur used for control of rust mites. As soils become more acid, the micronutrient elements leach out more readily, increasing deficiencies, and the basic phosphates which keep magnesium and calcium available are also leached. The nitrification process, by which ammonia is changed to the more readily available nitrate form, is inhibited in sandy soils when the pH gets below 5.5. Within the range of pH 5.5–6.0 there is maximum availability of mineral nutrients to citrus trees, as well as active nitrification, and it is very desirable that the grove soil be kept within this range of reaction. At pH values above 6.0 the micronutrient elements are held in the soil in forms increasingly unavailable to the trees, and the reaction should never be allowed to exceed pH 7.0. When high copper or manganese content makes iron unavailable, however, it is better to try to keep the reaction between pH 6.5 and 7.0. Control of soil reaction is normally accomplished by use of finely divided limestone.

The Citrus Fertilizer Program

A fertilizer program has as its objective making available for plant growth nutrient elements which are not obtainable under natural conditions. Some of these elements may be present in too low concentration for plants to remain alive long, some may be present and available in quantities which severely limit growth and yield, and some may be present in ample supply if only they were available to the plant. The fertilizer program must either augment the existing supply until it reaches adequate levels, or increase the availability of the supplies of unavailable elements. The citrus fertilizer program accomplishes both of these. It is designed not only with regard to present needs of the trees, to enable them to produce a current crop of desired size and quality, but also with regard to maintaining the trees in a satisfactory state of health and vigor for future production of good crops. The effectiveness of a fertilizer program, therefore, should never be judged by the results from following it for a single year, for these are largely influenced by previous programs. It is only by following a program for three or four years that a grower is in a position to assess its worth with confidence.

The citrus fertilizer program may be considered broadly as having three divisions: (1) the application of soil amendments, primarily for control of soil reaction; (2) the application of nutrient elements on the ground, or conventional fertilizing; and (3) the application of nutritional elements as foliar or nutritional sprays.

Soil Amendments and Soil Reaction

The term *soil amendments* covers materials which are incorporated into the soil for the purpose of improving soil characteristics and making them more favorable for plant growth. It properly includes organic matter mixed with the soil to improve its ability to hold water and nutrients in the case of sandy soils, and to increase aeration and tilth of clay soils. Organic sources of nitrogen, such as castor pomace and tankage, have some effect as amendments in addition to their primary function of supplying nitrogen, but the amounts added by their use in commercial fertilizers are so small that they can have only a negligible effect. Some growers have hauled into the grove such bulky and low-nutrient materials as water-hyacinths and meadow hay, but the cost of such practices usually far exceeds their value to the grove. Soil organic matter is usually increased best by growing and disking-in green manure crops.

The soil amendments of chief significance for Florida citrus groves are those which influence soil reaction, namely dolomitic and calcium limestones. Annually from 800 to 2,000 pounds of one of these materials may be required per acre to maintain the desired pH range on acid sands. Since each pound of sulfur applied in the pest control program requires slightly over 3 pounds of limestone to neutralize its effect on soil reaction, it is obvious that as little sulfur should be used as is consistent with good control of pests. The present trend toward control of rust mites by materials other than sulfur may change greatly the need for liming groves and make it possible to avoid the present fluctuations in soil acidity during the year.

Soil reaction should be checked annually, taking a number of samples in each grove to a depth of 6 inches from locations under the outer edge of the foliage canopy. Samples taken nearer the trunk are likely to give very low pH readings because of the drip of sulfur from leaves to ground, while samples taken in the row middles will be high because of no sulfur being involved to lower previous corrections of reaction. A composite sample combining individual samples from 20 trees is very desirable.

The amount of lime needed will depend not only on the pH reading but also on the texture and organic matter content of the soil. The lower the pH value, the more lime needed, and the finer the texture and the higher the organic matter content, the more the lime needed for a given pH. Under average conditions in sandy Florida soils, about 200 pounds of limestone are required to raise the pH value by 0.1 point, that is, from 5.0 to 5.1. If the very satisfactory practice is adopted of raising the pH by liming to 6.5 whenever it reaches 5.5, then 10 x 200, or 2,000, pounds (1 ton) of limestone will be needed to make this correction. Applications of lime may be made with equal satisfaction at any time of year; usually it is planned for a time when other grove work is slack. As indicated previously, dolomitic limestone is usually preferred because it supplies magnesium at the same time as it corrects acidity.

Application of Nutrient Elements on the Ground

Research has shown that the number and timing of fertilizer applications are of little importance provided a sufficient amount of needed nutrients in satisfactory balance with one another is applied annually, preferably during the months from October to June when heavy rains are less likely to leach away nutrients before the trees can take them in. The traditional practice has been to make three fertilizer applications a year: one in the autumn (October-November), one in the late winter (January-February), and one in early summer (May-June). The first of these assures the tree of a store of nutrients to replenish what is lost in fruit harvested and leaching by summer rains; the second is immediately important for fruit set and new vegetative growth; and the third provides for rapidly developing fruit and for the summer growth flush. This practice is eminently safe and satisfactory, but there is expense for labor as well as for fertilizer at each application, and fewer applications would save money if the trees thrived as well. Many growers have found that they can get along very satisfactorily by making only two applications per year, in October-November and in April-May, supplying the same total amount of nutrients as would be supplied in three applications usually. There is good research basis for considering it probable that a single annual application made in April-May would be equally satisfactory, for citrus trees have shown an unexpected ability to absorb and store nutrients for future use, but this is not firmly enough established to be recommended as a safe practice yet.

The needed nutrient elements may be provided from many source materials and in various types of mixtures. Regardless of the sources and combinations, the amounts and ratios of the elements are the items of real importance for a successful program. The commercial grower must vary this program to take account of variations in stock-scion combinations, in environmental conditions, in cost of fertilizer materials, and in market demands and prices.

Fertilizer recommendations for commercial groves of bearing trees over 10 years old are based on the expected yield per tree, as estimated from the known yields of the past few years. The average orange tree, planted 25 feet by 25 feet, increases in yield about ⅕ box per year up to about 35 years of age and decreases at the same rate thereafter. Only a very small allowance need be made, therefore, for the fertilizer necessary to take care of the larger yield of the tree to be expected this year over the yield harvested last year.

The basis of fertilizer application currently favored by research men is to supply 0.4 lb. of nitrogen for each expected box of oranges, mandarins, lemons, limes, or tangelos, while 0.3 lb. per box is sufficient for grapefruit. Other elements are added in a fixed ratio to nitrogen. Potash is usually supplied in equal quantity with nitrogen, magnesium at about half this rate if dolomitic limestone is used to control pH, or nearly the same rate if calcic limestone is used. On young bearing groves under 10 years old, phosphoric acid is included at ¼ the nitrogen amount, manganese at 1/16, copper at 1/30, and boron at 1/100 this amount. On groves over 10 years but under 20 years old, phosphorus may be reduced by half and the three micronutrient elements eliminated unless deficiency symptoms indicate their need. After the 20th year no further application of phosphorus is likely to be profitable. Iron in the chelate form is applied either mixed in fertilizer or as a separate application when deficiency symptoms show its need.

Usually fertilizer mixtures containing all of the needed elements are used, and by convention the successive numbers represent the per cent of available nutrients in the order N-P_2O_5-K_2O-MgO-MnO-CuO. Formerly mixtures with 4 per cent nitrogen were common, but currently the minimum percentage of nitrogen is usually 6, and the mixture may contain 8, 10, or even 15 per cent, because of decreased cost of applying smaller quantities of higher analysis.

The following chart will illustrate the common fertilizer analysis for the different age groups just mentioned:

Trees under 10 years	Trees 10 to 20 years	Trees over 20 years
6-2-6-3-0.4-0.2-0.1*	6-1-6-3	6-0-6-3
8-2-8-3-0.5-0.25-0.1*	8-1-8-3	8-0-8-3
10-3-10-4-0.6-0.3-0.1*	10-1-10-4	10-0-10-4

*B_2O_3. In groves which receive applications of dolomite, the amount of MgO can be decreased.

Up to the age of 10 years, a citrus tree is rapidly increasing in size, and fertilizer applications must take account of the vegetative growth as well as of crop. For the young bearing tree, therefore, fertilizer recommendations are better based on age than on yield. By the time the tree is 10 years old, vegetative growth has usually slowed down to a more steady pace and there are several years of fruit production on which to base yield expectations. Even with mature bearing trees, however, there may be situations where the owner has no knowledge of previous production by a newly acquired grove and so has no basis for calculating fertilizer needs. In such cases he may safely use age as a guide until he has yield records for two or three years. The following table shows the average yields to be expected from orange trees on high sandy soil, spaced 25 feet apart, at various ages, and the amount of fertilizer to be used.

Age of Tree	Approximate Yield	Pounds of Fertilizer per Tree at Each of Three Applications per Year for 6% N	8%N	10%N
4 years	½ boxes	5	4	3
5	1¼	6	4	4
6	2	7	5	4
7	2½	7	5	4
8	3	8	6	5
9	3½	8	6	5
10	4	9	7	5
15	5	11	8	7
20	6	13	10	8
25	7	15	11	9
30	8	17	13	11
35	9	19	15	12
40	8	17	13	11
45	7	15	11	9
50	6	13	10	8

The amounts per tree and per acre at each of three applications per year for trees over 10 years old for various estimated yields to be expected are shown for these three common analyses in the tables at the top of page 169.

If all fertilizer for the year is put on in two applications instead of three, then the amount at each application should be increased by half. For grapefruit trees, apply ¾ of the above amounts for any given yield in boxes as shown. In commercial groves the fertilizer is usually spread by machine from trunk to trunk across the full width of the middles.

Application of Nutrient Elements in Sprays

Usually when spraying citrus trees is mentioned, it is pest control that is in mind; but spraying to provide certain micronutrient elements is standard practice in citrus culture. These elements may be applied separately but are usually incorporated in sprays for pest control for reasons of economy. Nutritional sprays for Florida citrus trees date

Boxes per Tree	Pounds per Tree at Each of Three Applications of Mixed Fertilizer with 6%N	8%N	10%N
2	4	3	3
3	7	5	4
4	9	6	5
5	11	8	6
6	13	10	8
7	15	11	9
8	17	13	10
9	19	15	12
10	22	16	13

Boxes per Acre	Pounds per Acre at Each of Three Applications of Mixed Fertilizer with 6%N	8%N	10%N
200	445	335	265
250	555	415	330
300	665	500	400
350	775	585	465
400	890	670	530
450	1000	750	600
500	1110	835	665
550	1220	915	730
600	1330	1000	800
650	1440	1085	865
700	1550	1170	930
750	1660	1250	1000

from around 1934, when zinc was found to be required and to be poorly available from ground applications but readily so in sprays. Other micronutrient elements are occasionally applied as sprays.

Nutritional sprays are costly to apply by themselves, and the residues which are left by them on the foliage contribute to increase in populations of insects and mites by shielding them somewhat from their enemies. While these sprays are often beneficial to the tree, or even necessary to its well-being, undesirable or harmful plant reactions may result from their indiscriminate use. They should be used only when benefits can definitely be expected. The two situations which always warrant their use to correct observed deficiencies are (1) when applications to the soil are not effective, and (2) when a more immediate response is considered desirable than can be obtained by soil applications. Nutritional sprays are satisfactory only for supplying elements needed in very small amounts—the micronutrient elements, or for supplying magnesium (as magnesium nitrate) to trees on alkaline soils. They may be *corrective* sprays, aimed at correction of an observed deficiency condition, or *maintenance* sprays, intended to hold the nutrient level above the point of observable deficiency by regular application of smaller amounts than were needed to correct the deficiency.

Zinc compounds are absorbed to only a very small extent by citrus trees from soil applications, and as this element is very commonly deficient in Florida soils, it is regularly applied as a nutritional spray. For such maintenance spraying, 2½ pounds of metallic zinc per 500 gallons of water satisfies the annual requirement. This amount of zinc may be provided by 7½ pounds of zinc sulfate plus 2½ pounds of hydrated lime, or by 5 pounds of a neutral zinc compound (55% Zn). The neutral form will leave less residue on the leaves to affect insect and mite numbers, and so is preferable to the sulfate plus lime. The spray application may be made either during the dormant period or shortly after blooming, and need reach only the outside of the foliage canopy. If zinc deficiency symptoms have developed in the grove before any nutritional spray has been used, a corrective spray will be needed. This should contain twice as much zinc as the maintenance spray and usually need be used only one season.

Fig. 8. Loading a bulk-fertilizer distributor for efficient broadcast application in citrus groves.

Courtesy Haines City Citrus Growers Association

Yellow spot, due to molybdenum deficiency, seems to appear more commonly on trees on grapefruit stock and is noted first in summer as small, water-soaked areas on the leaves. Mild symptoms can be corrected by spray containing 5 ounces of sodium molybdate per 500 gallons of water, while severe symptoms require double this amount. Only the outside of the canopy needs to be covered. Spraying is done only when deficiencies are seen.

Manganese and boron are often poorly available from ground applications on alkaline soils and their deficiencies must then be corrected by sprays. The same is true of copper if no fungicidal copper sprays are employed. Manganese and copper are used at the rate of 15 pounds of metallic sulfate plus necessary hydrated lime per 500 gallons of water, or an amount of a neutral compound which is equal in metallic content to the above. As pointed out in discussing zinc, the smaller amount of residue left by neutral compounds of metals results in less of a pest problem. Boron sprays regularly contain 1.25 pounds of soluble borate (58–66% B_2O_3) in 500 gallons of water. In all cases of nutritional sprays, avoid excess applications.

In summary, the complete nutritional program for commercial citrus production must include (1) soil amendments of liming materials to adjust the soil reaction to the pH range of 5.5 to 7.0 for acid soils; (2) application on the ground of sufficient amounts in proper proportions of those elements not naturally present or available in the soil, through a two- to three-application fertilizer program; and (3) nutritional sprays as needed to supply elements which cannot be furnished by soil applications.

9. The Citrus Spray Program

THE commercial production of citrus fruits requires varying amounts of spraying and dusting operations. Dusting can replace spraying in only a few cases; this discussion will therefore be concerned primarily with spraying, with reference to dusting where it is a suitable alternative. The sand hill groves of central Florida require the greatest amount of spraying, with costs of this operation often amounting to nearly half of total production costs. In other areas the spray programs are often less extensive; yet in no commercial area of the state can a proper spray program be overlooked. The "no spray" program of the early citrus growers is no longer valid. These differences are due in part to the activity of parasitic and predatory insects and mites, of snails, and of beneficial fungi, and in part to differences in the influence of environments on the pests themselves.

The commercial spray programs for citrus fruits have become more complex with passing years, due to the increasing age of trees and the increased density of plantings. These programs include application of insecticides, miticides (or acaricides), and fungicides, and of physiological sprays for nutrition, fruit maturity, and growth regulation. Out of these the commercial fruit grower must develop a spray program for his particular grove, based on the number, kinds, amounts, and importance of the various items needing control. He must be concerned with materials, dosages, coverages, timing, and the precautions needed to avoid damage to plants and people. Injury to plants by materials offered on the commercial market for spray application is kept to a minimum by close adherence to current recommendations, although even then the commercial grove may at times receive some damage due to conditions beyond the grower's control. Many of the newer materials are hazardous to use from the standpoint of the operator; therefore, instructions on the containers must *always* be read, understood, and followed carefully. Residues on fruit at time of harvest are sometimes of toxic nature; EPA tolerances on citrus fruits have been determined for many of these materials and are stated in parts per million. Along with tolerances are statements with respect to the kinds of fruit for which the materials are approved and the minimum days before harvest that they may be applied. The operator may obtain such information from the Florida Citrus Spray and Dust Schedule published annually by the Florida Cooperative Extension Service. Regulations of the EPA must be respected. The grower should read the label for instructions and limitations, apply the materials properly, store unused material safely, and dispose of empty containers out of reach of people and animals and to avoid contamination of environment.

Since a grower cannot afford the extra expense involved in making separate spray applications for each of several problems which may be present at the same time, he must also be concerned with compatibility of materials; that is, with what spray materials can be combined for application. In general, incompatibility may take two forms: (1) combining two or more materials in one spray may result in an increase in injury to the tree, as is true of combining oil with sulfur (indeed, oil and sulfur cannot be safely applied within a month of each other); (2) the effectiveness of one or more of the materials may be decreased by mixing them, as is the case when chlorobenzilate is mixed with any alkaline material.

In the following discussion some references will be made to minor problems, but the major factors related to the spray program will be emphasized.

Pest Control Sprays

Scales, Whiteflies, and Mealybugs

This group of closely related insects comprises a larger number of active pests of citrus trees than any other in the state. Although individual species sometimes may require particular control measures, usually control for one member of this group will control other members also. Therefore, a single application of spray may be effective for several different pests.

Purple Scale—This armored scale was for many years the most destructive insect pest of citrus fruits in all areas of citrus in Florida. Until the last decade it has almost invariably required regular spraying for its control, but this situation has changed greatly since the introduction of a new parasite (*Aphytis lepidosaphes* Compere) in 1958. With its establishment throughout the citrus belt during 1958, purple scale had been reduced to such low levels of population that little or no spraying is necessary. Weather conditions appear to have little relation to the effectiveness of this parasite, but when sulfur or lime-sulfur is used repeatedly, their harmful effect on the Aphytis will probably make it necessary to include a scalecide in the spray program to control purple scale. Since it is difficult to evaluate the long-term value of this friendly parasite, a full discussion of purple scale and its control is here given; but it is fondly hoped that this major pest will continue to prove as unimportant as cottony-cushion scale.

FIG. 9. An efficient spray operation with use of an air-blast sprayer.
Courtesy Haines City Citrus Growers Association

The slender, dark purple, male scale is usually found on the upper surface of leaves, while the much larger female scale, a little over 1 mm. long and club-shaped and brownish, mostly prefers the lower sides of leaves. These scales often congregate in groups, and by their sucking of cell sap cause yellow-brown spots on the leaves. Where they have fed on fruits, green spots persist even after the rind develops its mature color. They also attack twigs and even limbs. The long scale, very similar to purple scale but longer, is often found mixed in these colonies.

The damage caused by purple scale is of several kinds. The sucking of food from leaves impairs seriously their ability to make and transmit food to other parts of the plant, and the heavier the scale population the greater this impairment. In cases of large populations, the leaves may be so weakened that they fall off, and for lack of food both immature and mature fruit may be shed. Dropping of fruit may also result from feeding of scales on the fruit around the stem, especially in early fall. Many twigs are killed by feeding of this scale, especially defoliated twigs, and this not only increases cost of pruning but also provides a place for the fungus causing melanose to multiply. With loss of leaves and twigs, or even branches, there is a decreased bearing surface for the following year and less production of flowers on remaining twigs. The persistent green spots on the fruits lower the grade of the fruit, and the scales themselves are difficult to remove from fruit going through the packinghouse. These serious damages, formerly so commonplace, will rarely be seen today; yet, unfortunately, small infestations of purple scale (as with long and chaff scales) may leave the green spots on the mature fruit.

Control of purple scale has long been effected by the use of oil emulsions or miscible oils at the rate of 1.3 per cent of actual oil (e.g., 7.0 gallons of a 92 per cent oil emulsion in 500 gallons of spray). It is usually applied during early summer to give a thorough coverage of both the inside and the outside of the canopy; however, in the Indian River area dormant or postbloom applications may be very effective. Oil sprays must not be applied within a month of applying sulfur, nor should they be used during either very dry or very cold weather. They may cause a rind injury known as "oil blotch" if applied to fruit from ¾ to 1½ inches in diameter, and if used after the end of July they may delay coloring and maturing of fruit. Late fall application makes citrus trees more susceptible to injury by cold and so may decrease the crop of the following year. Even in the absence of killing cold such late oil sprays may affect the next year's crop unfavorably. They should not be applied twice in the same season, as this increases their undesirable effect on tree activities. Oil sprays are free from hazard, however, in their effect on people.

In commercial practice the organic phosphate parathion controls purple scale well at dosages of 1.25 pints of parathion 4 liquid (or equivalent) per 500 gallons. This material can be used equally well at any time of year, so far as injury to the tree or fruit is concerned, unless very tender new shoots are present, but is very dangerous to the user unless proper safety precautions are carefully observed. It is even somewhat hazardous for people to walk in a citrus grove soon after it has been sprayed with parathion; current regulations of EPA concerning grove re-entry should be understood. It causes headache, nausea, and sometimes death to those who come in contact with it carelessly, though it is safe enough when used carefully. Malathion, another organic phosphate, is much less dangerous to the user, though still requiring caution, but must be used in higher concentrations (5 pints of malathion 5 liquid per 500 gallons) and even then does not give quite such effective control. The homeowner should never use parathion but may employ malathion with care. Coverage of trees with either must be very thorough for satisfactory control.

Florida Red Scale—Although not distributed as generally over the citrus belt as purple scale, Florida red scale formerly required equal effort for control and sometimes, especially in the southern part of the state, superseded purple scale as the major insect pest in some groves. After mild winters it was apt to occur in unusually heavy infestations. The armor of this scale is circular with a prominent central nipple, is up to 1.5 mm. across, and varies in color from reddish-brown to reddish-purple.

The damage caused by Florida red is similar to that of purple scale, with two important differences. It does not feed on twigs, and the spots where it has fed on the fruit become yellow prematurely instead of failing ever to yellow. Control measures are exactly the same as for purple scale. However, due to the activity of parasites, the early scourges of this pest are no longer experienced.

Citrus Snow Scale—This scale has been in Florida for many years and was a serious problem in parts of Orange and Seminole counties. The name is derived from the fact that the clustered white male scales look like driven snow on the trunks or limbs, where they are mostly found. Since 1969, it has spread over the state and by 1971 had become a serious pest, considered today to be the most serious of the armored scale insects. Feeding on the trunks and limbs, chiefly, heavy populations spread onto the leaves and fruit. The homeowner will find a thorough application of an oil emulsion (see above) safe and satisfactory. For control in commercial groves, 2 sprays per year (following bloom period and during the summer) of phosphatic insecticides thoroughly applied, and with complete coverage of all wood, are required. Obtain current recommendations *with precautions* for use of these hazardous materials. Research with respect to possible parasites of the citrus snow scale is now in progress.

Glover Scale—This relative of purple scale, which is often referred to as the long scale, does the same damage and is controlled similarly. This scale has been in the state since the introduction of citrus but was found in only small numbers; it was thought to be held in check by parasitic activity. Surprisingly enough, as purple scale became less important, Glover scale populations increased. It can be distinguished from purple scale by its longer, narrower armor, particularly of the female.

Yellow Scale—With a round armor much like that of the Florida red scale but lighter in color, this scale was for many years a commercial pest in Pinellas County. No eggs will be found under the armor since the females give birth to living young. Like the Florida red scale, it attacks only leaves and fruit. In recent years, this scale has spread throughout the state and at times has demanded control with oil emulsions or other insecticides.

Chaff Scale—These scales are brownish-gray, nearly circular, often clustered densely, especially on branches, like overlapping pieces of wheat chaff. The most distinguishing feature of this scale is the purple color of the female, the eggs, and the crawlers. With heavy infestations the leaves and fruit develop considerable populations of this scale, and the persistent green spots left on the fruit by its feeding are particularly serious on tangerines. The chaff scale appears to hold its own as a pest in various groves throughout the state.

Other Scale Insects—While the above-mentioned six scales belong to the armored scale group whose representatives are major pests in all parts of the citrus world, there are other scale pests which may, at times, be injurious.

The cottony-cushion scale belongs to an entirely different family. The adult female is conspicuous because of its greatly expanded, fluted, white cottony abdomen filled with red eggs. It is rarely a serious problem in bearing groves because the Vedalia ladybeetle, introduced to Florida from Australia via California in 1899, keeps the populations low. Control of this scale by the Vedalia is the classic example of biological control. In the nursery, however, it may require control measures at times. Oil sprays cannot penetrate the cotton-like mass and therefore are ineffective for control. Parathion or malathion can be used in commercial practice but should not be used around the home. The homeowner will find an old toothbrush quite satisfactory for brushing off the scale.

Another family, the soft scales, whose apparent armor is part of the actual insect body, includes the black scale, brown soft scale, and green scale. These rarely appear in damaging numbers, unless parasite activity is retarded, although black scale has required control measures with oil emulsion or other insecticides at times. Ridges on the upper part of the body of the female black scale portray the letter "H."

Citrus Mealybug—Usually not a pest requiring control, this insect may at times become numerous enough to cause serious injury. It is most readily recognized by its white masses, enveloped in a cottony frass, which are seen on tree trunks in very early spring. In heavy infestations this mealybug causes dropping of fruit and lowering of grade, especially with grapefruit, because of the hard lumps on the fruit rind, which become yellow prematurely. Oil emulsions are not effective in control since they will not penetrate the cotton-like mass. In commercial operations, but not on the homesite, 2.5 pints of parathion 4 liquid (or its equivalent) per 500 gallons can be used. Safer, but still to be used with caution, is malathion 5 liquid at a rate of 6 pints per 500 gallons. An old toothbrush will remove excessive numbers from a few trees around the house. Any control measure is best applied before blooming or immediately thereafter, before the crawlers have become established under the calyx (button) of the fruit.

Whiteflies—The common citrus whitefly, which lays yellow eggs, and the cloudy-winged whitefly, whose eggs are black, are both serious pests at times. Often they occur together, but the former is more abundant in the northern area and the latter in the central and southern areas. The woolly whitefly is widely distributed but rarely troublesome. The name whitefly comes from the adult form, which is a small, two-winged, flying, white insect and is harmless in itself. The damage is done by the feeding of the larval stages (nymphs), which are much like scale crawlers and suck food similarly from tender leaves, especially those of vigorous sprouts. They are found mostly on the lower surface of leaves, rarely attacking the fruit. These whitefly larvae, along with aphids, mealybugs, and soft scales, produce a sweet secretion, called honeydew, which serves as food for the sooty mold fungus, often seen as a black film coating leaves or fruit. Besides making fruits unattractive, this coating delays the natural loss of green color as the fruit matures and serves as a protection for various other pests.

Many years ago whiteflies were more abundant and harder to control than they are now, and constituted a serious problem for growers.

Natural enemies, especially the friendly fungi, and the sprays for control of scale insects have long since reduced populations to a low level. Control is easily obtained with the same materials used for scales, and usually sprays applied for the latter take care of whiteflies also, so that they rarely need special control.

Other Insect Pests

Aphids—The aphids, or plant-lice, feed upon succulent growth, especially very young shoots only partly developed. Several species of aphids are found generally throughout the citrus belt. Although winged forms may occasionally be seen, aphids are usually observed as colonies of wingless forms densely clustered on tender shoots. The young leaves from which they suck food become curled inward toward the feeding area, and remain permanently dwarfed and twisted. The green citrus aphid is perhaps the most destructive species, but the melon and black citrus aphids do very similar injury to new growth. All these species are able to transmit the tristeza virus disease.

Control of aphids is usually not necessary in groves of mature orange, grapefruit, or tangerine trees, but may be needed with 'Temples'. Trees which produce a succession of succulent growth flushes, trees which have unusually vigorous and succulent growth because of heavy pruning, or trees whose foliage is valued highly may warrant spray applications. Young trees, whose foliage is worth so much to future crops, often require control measures for aphids.

Aphids multiply rapidly and do their damage quickly. Any control measures must be anticipated in order to kill the pests before deformation of leaves has occurred and, for ideal control, repeat applications are often required. As vectors of tristeza, all three species must be considered in areas of active spread of this virus. There is no safe aphicide for use around the home; however, a soap solution, 1 pound of Ivory soap to 6 gallons of water, properly sprayed on the tender foliage will do some good. Systox 2 liquid at 5 pints per 500 gallons has been used for many years in commercial practice with proper precautions. Other aphicides which require very careful handling are now approved for commercial usage but, like Systox, are very toxic to humans.

Plant Bugs—The true bugs found as pests in citrus groves include the citron bug, which feeds on the seeds of the wild citron (related to watermelon) found on the sand hills of central Florida, and the leaf-footed plant bug and the stinkbugs, which feed on the immature pods of crotalarias and other legumes. After their usual food supplies have dwindled in the fall, these bugs often fly to nearby trees and feed on mature citrus fruits, causing them to drop. Since they puncture the rind with their piercing-sucking mouth parts, it is almost impossible to see where this feeding was done unless the rind color is slightly changed at that point. These pests have declined in importance in the last decade.

The best control measure is to cut down the cover crop in the grove in September, before citrus fruits have become attractive to these bugs or while they are still in an immature, wingless stage. Formerly they were very difficult to kill by spraying, but there are now several chemical compounds which can be sprayed or dusted on trees and cover crops for their control. Most of these chemicals are also very toxic to humans and must be handled carefully. Current recommendations should be obtained if chemical control becomes necessary.

Grasshoppers—These pests breed in outlying fields, or even in citrus groves at times, and may develop to such numbers as to become very harmful when the grasses on which they usually feed are eliminated by cultivation or are killed by drought or frost in late autumn. Then they feed on citrus leaves and may destroy practically all the foliage in a grove in a short time. Two species are common in Florida, the large American or bird grasshopper, which is sometimes a very serious pest, and the even larger but less active lubber grasshopper, which is only of minor importance. If they are developing in fields adjacent to a grove, chemical control may be necessary after the grove cover crop has been chopped or disked. When the infestation develops within the grove, however, proper management of the cover crop is very helpful in control. Clean cultivation during the early spring (February to May) and after the middle of August will reduce grasshopper infestations by preventing laying of eggs in the grove soil. When chemical control seems necessary current recommendations should be obtained. As the native fields and woodlands of the state are brought under cultivation or used for urban development, grasshoppers are losing many of their breeding grounds. The large infestations of former years will probably not occur again.

Root Weevils—Several species of weevils (often referred to as beetles) cause serious damage to citrus trees, particularly in the larval stage which feeds upon the roots. The adult stage feeds upon the leaves but damage is not serious; feeding is often so slight that the insect would hardly ingest sufficient material to kill the leaf if an insecticide were applied to the aerial parts of the tree. The grayish-brown Fuller rose beetle has been known for many years in Florida. In 1952, the white, legless larvae were first found feeding destructively on the roots of citrus trees in the Indian River area; since then it has been reported from the other citrus-growing areas of the state. The adult causes some injury by feeding on the leaves, making characteristic notches in the margins. The citrus root weevil, known as the blue-green beetle in the Indian River area, is a serious problem there and may be found in other citrus regions. The sugarcane rootstalk borer weevil was inadvertently introduced into Florida and found near Apopka (Orange County) in 1964. This weevil has a wide range of host plants and constitutes a serious threat to the citrus industry. The Division of Plant Industry is carrying on extensive programs to limit its distribution while rearing parasites which effectively control the pest in Puerto Rico.

Control of these weevils and others with similar habits cannot be obtained by application of sprays to the trees because of the small amount of feeding of the adults. Sprays applied to the soil would be suggested but it is cheaper and most effective to apply insecticides at the same time with fertilizers. Aldrin and dieldrin are satisfactory materials for control and give best results when 2½ pounds per acre of one of them is mixed with fertilizer for both fall and spring applications. This method of usage is the only type legal today. Since these materials are chlorinated hydrocarbons, that are under the regulation of, and may be withdrawn from the market by, the Environmental Protection Agency (EPA), growers should obtain current recommendations prior to use.

Miscellaneous Insects—Many other insects may be seen in citrus groves, some of which may at times cause a little damage, others of which are actually beneficial (see later), and still others merely like the grove environment; these three groups become a part of the biotic environment. By observation the grower will come to recognize some of these and to understand their place in the grove complex. Among them will be flies of several kinds, and these may cause the grower to worry lest they should be Mediterranean fruit flies. These latter are very serious citrus pests in some parts of the world and while they have been successfully eradicated from Florida twice, at great expense, after being accidentally introduced, another introduction may start an infestation again. The grower does well to be constantly on guard against this event, and should send any suspicious specimens to the Division of Plant Industry at Gainesville for identification.

Mites

Mites are related to spiders, often spinning webs, and several kinds of them make trouble for the citrus grower. They differ from insects most obviously in having 8 legs normally, whereas insects have 6, but they are so small that the grower is not likely to try leg counting. By far the most important is the rust mite.

Rust Mite—This was one of the first pests to be noticed by early citrus growers in Florida because of its causing oranges to be "russeted" or grapefruit, lemons, and limes to show the silvering of rind called "sharkskin." It also causes rind pitting of all these fruits and may cause "firing," a drying of leaves while still attached to the twigs, and defoliation on a large scale, especially in the dry winter months. It has also been associated with the presence of "greasy spot," a disease discussed later in this chapter. The presence of rust mites is difficult to detect until they have demonstrated it by their characteristic injuries, because they are exceedingly small—hardly 1/200 inch long. The yellow, conical, segmented body, broadened at the front end, has only 4 legs (the other 4 having been lost) and can be detected only with the aid of a good hand lens. Rust mites feed on leaves, fruit, and green twigs, and are capable of causing serious injury. While the population peak is usually reached in midsummer, and fruit russetting occurs in the period from postbloom to September, this mite should be kept under control throughout the year. Damages to the tree occur principally in the winter dry period. The rust mite has shown a continuous population increase since 1964, and is considered the most important and widespread citrus pest in Florida.

For many years the standard (and only) material for control of rust mites was sulfur. Wettable sulfur can be used at the rate of 25 pounds per 500 gallons in the postbloom period and as a supplemental spray between main sprays when control is required. Sulfur dust, at 0.5 to 1.5 pounds per tree, according to size, is effective for short period control, especially with thorough applications made at night when trees are wet with dew in periods of the year of low rainfall. The sulfur is gradually washed off the tree and produces an undesirable increase in acidity in the soil; furthermore, continued use has been followed by increases in spider mite and armored scale populations. Its use should be alternated with that of other miticides which do not produce these effects. Zineb, at 5 pounds of 75 per cent powder per 500 gallons, was the first substitute for sulfur but today its effectiveness is questioned. Chlorobenzilate 4 liquid, at 1.25 pints per 500 gallons, has had widespread usage. Other miticides, including those which also control the spider mites, are now available. These miticides should be applied when the population has reached 10 per cent, prior to damages, and in the standard spray schedule require several applications per year.

Purple Mite—Citrus red mite is the approved name for this pest, but the name purple mite is unfortunately firmly established in Florida. It is one of a group of mites often called "red spiders," and is about 1/70 inch long, rose-red to purple in color, with the typical 8 legs except in the first stage, when it has only 6. The tiny, round, red eggs can be seen with a hand lens along the midrib near the base of leaves. This mite attacks leaves, fruits, and green twigs, but the principal feeding is done on the upper leaf surface.

The purple mite has been in Florida for many years, sometimes producing damages which were ascribed to other causes, but damages began to increase in severity about 1935 until it became one of the major citrus pests. However, its importance has declined in recent years. The feeding causes leaves to become grayish and produces mesophyll collapse, characterized by translucent, or necrotic, areas of the leaf due to death of mesophyll cells. Firing, drying, and browning of leaves while still attached to the twigs, and defoliation, often occur in the winter and spring as a result of heavy infestations. The mite also feeds on the green wood and fruit; it causes a scratched appearance on the green rind of the fruit but the final grade is not impaired.

Prevention is always better than cure, and spraying in the fall when 10 per cent of the shoots show infestation will prevent this mite from reaching damaging numbers during the winter months when most tree damages would occur. Control with sulfur is not effective; in fact, this material may actually increase populations. Oil emulsion, at 0.5 per cent actual oil, kills both the eggs and the hatched mites, but its use is limited to the summer months since fall and winter applications produce the undesirable tree effects discussed under purple scale. When oil emulsion is applied for scale control in postbloom and summer periods, control of spider mites is also obtained. The combination miticides, such as Ethion, Trithion, Delnav, and Kelthane, which kill both spider mites and rust mites, can be used in the fall and winter months. These should be used at recommended dosages and with proper precautions because of their hazardous nature. All of the miticides discussed must be used with care to avoid leaving excessive toxic residues on the fruit. It is always advisable to obtain current manufacturer's recommendations.

Texas Citrus Mite—This mite was found for the first time in Florida in 1951 in Brevard County and had become generally distributed over the state and a major pest by 1966. Under a good lens it can be seen to be tan to brownish-green in color, with greenish spots along the sides. Like purple mite, it is found chiefly on the upper side of leaves, and its tiny, flat, discoid eggs, light tan in color, are laid along the midrib. It causes the same sort of injury as purple mite and is controlled by the same materials.

Six-Spotted Mite—This long-time pest in Florida is a heavy-bodied, yellow mite with six dark spots on its back, which lives in colonies only on the lower surface of leaves and is very widely distributed throughout the citrus area. Its appearance in individual groves is sporadic, requiring vigilance on the part of the grower. During the summer period it cannot be found in the groves. Like purple mite, this pest increases in numbers rapidly during the winter and spring months, if not controlled. It prefers grapefruit varieties as hosts, but may also be found on other kinds of citrus trees. It seems to prefer areas in the grove of low humidity and is often most abundant in trees next to a road. Damage consists chiefly in the greatly decreased ability of leaves to make food because of its feeding, but leaves may drop in severe cases. Heavily infested leaves are characteristically crinkled, with large yellowed areas. Dropping of fruit may be an indirect result of impaired leaf activity.

Control may be obtained with any of the materials recommended for purple mite, but the time for most effective control is just before the spring flush of growth, certainly before the leaves have become chlorotic. The spraying must be done in any case so that the undersides of all leaves are wetted, since only those mites actually wet by the spray will be killed.

Other Mites—The leprosis mite, which causes leprosis of branches (Florida scaly bark) and fruits (nailhead rust) of orange trees, was considered a serious pest in the first quarter of this century. In recent years it has rarely caused much damage, probably because it is held in check by the constant use of sulfur for rust mites. Because of uncertainty as to its control by other miticides, an annual application of sulfur (50 pounds of wettable per 500 gallons) is recommended where it is a problem. The application is best made in the late dormant period.

The broad mite feeds on young fruit and immature leaves. While it is sometimes observed, particularly along the East Coast, by its causing "sharkskin" blemish on fruit and rolling of leaf margins, it rarely causes much damage. The spray for leprosis mite will control it also.

The citrus bud mite was first recorded in Florida in 1959 from Dade County. How serious this pest, which distorts tender new shoots, will prove remains to be seen.

Predacious mites, discussed later, are sometimes mistaken by growers for mite pests, resulting in unnecessary spraying. They cause no plant injury.

Diseases

Diseases may be caused by bacteria, fungi, algae, virus, or nematodes. The latter two groups cannot be controlled by sprays or dusts and so are not included in this discussion of the Spray Program. They are dealt with in Chapter 7.

Citrus Canker—This is the only bacterial disease of citrus trees which has been found in Florida. It was first found in 1913, following its accidental introduction in 1910. An intensive program of eradication was instituted at once, resulting in complete success within a few years. This eradication program was the primary factor in establishing the need for the services of a Division of Plant Industry (formerly State Plant Board). Citrus canker has not been found in the state since 1926, but might be introduced again easily from other parts of the world. Plant inspectors at entry ports watch vigilantly to prevent this.

Melanose—The fungus causing this disease produces its spores only on dead twigs but attacks only young leaves, fruit, and twigs. The chief damage of melanose is the lowered grade of infected fruit. It appears as brownish, raised lesions on affected parts, but the same fungus also causes Phomopsis stem-end rot of mature fruit after harvesting. Melanose costs more for its control than any other production disease, for fruit of good grade can only be assured by regular sprays for its control. Pruning of dead wood is an important aid to control, but is not adequate by itself.

Neutral copper compounds are now used extensively, at the rate of 3.75 pounds of metallic copper per 500 gallons. Standard practice is to supply this as a slow, outside cover spray from one to three weeks after petal fall, when the tiny fruit can first be seen to have set. It is not safe to delay this spray as much in a very rainy bloom period as in a drier one. If melanose was severe the previous year, a second application of the same spray should follow a month later. Delaying the first spray more than three weeks after bloom may intensify grade lowering of melanose lesions by adding copper injury (star melanose).

Postharvest decay due to Phomopsis stem-end rot, as well as Diplodia stem-end rot and green mold, is controlled by applying a fungicide to the fruit after harvest. However, recent work with Benlate, a fungicide, has significantly controlled these types of decay when applications are made to the trees in the fall period.

Scab—'Temples', satsumas, lemons, sour oranges, certain varieties of grapefruit and tangelos, and young 'Murcott' trees are often attacked by this disease, caused by a fungus which grows in green tissues and causes unsightly warts on fruits and distorted leaves and twigs. On fruit early scab lesions differ from those of melanose in being much lighter in color, larger in size, and more elevated; fruit and leaves are often much deformed by early infection with scab, whereas this is not true of melanose on fruit. Except for injury to nursery stock of sour orange and 'Rough' lemon, the principal damage by scab is lowered fruit grade, as with melanose.

Control of scab is effected by neutral copper sprays also, but dosage and timing are different from those for melanose. Content of metallic copper should be 5 to 7 pounds per 500 gallons and the critical period for application is just before the spring flush appears. A slow outside cover is satisfactory. When scab was serious the previous season, a second application may be needed at petal fall. Ferbam, at 7.5 pounds of 76 per cent powder or its equivalent, is more effective than copper for this second application at ⅔ petal fall. In 1972 a new material and new principle were introduced for scab and melanose control. Difolatan 4F, at 4 to 5 gallons per 500 gallons, was recommended in a single application as a delayed dormant spray. It should only be used in groves with overhead irrigation to further distribute the fungicide. Rind blemishes may occur which prevents the use of the material on mature fruit or in the postbloom on fruit destined for the fresh-fruit market; label recommendations should be followed carefully.

Greasy Spot—This disease, which is seen as greasy-appearing spots scattered irregularly over the leaf surface, may be observed at any time of year, but is often more prominent on fall flush leaves. It is caused by a fungus, and can be controlled by applications of spray about a month after any flush has developed. Neutral copper compounds at 1.25 to 2.50 pounds of metallic copper per 500 gallons will control this disease satisfactorily, as will summer sprays of oil for scale control. Use of copper spray is not recommended during the summer on fruit which is intended for the retail market, since copper tends to darken any lesions on the rind, thus lowering fruit grade. This condition, "star melanose," is particularly severe when copper is applied to fruit with melanose lesions in summer.

Brown Rot—In coastal areas especially, but also in other areas in years of high rainfall, early and midseason citrus varieties are apt to show rotting of fruit on the tree in early autumn due to infection by the fungus causing foot rot. Damage can be much reduced by a spray of 2 pounds of metallic copper in 500 gallons, applied in early September to those groves with a past history of the disease. It is not a common problem. Since the source of infection is at ground level, the spray should be directed primarily at spores on the ground and fruit in the lower part of the canopy. Such cultural practices as pruning low-hanging branches and mowing or disking the cover crop will also be helpful in control by permitting better circulation of air and lowering humidity.

Other Diseases—Many other diseases attack citrus trees from the nursery years, when damping-off of young seedlings is a hazard, to those of advancing age in the grove, when dieback or root rots afflict them. Red-alga spot is the only algal disease and affects seriously only limes and lemons, which may need copper sprays at times for its control. Damping-off and anthracnose in the nursery may also require spraying. Otherwise control of the other citrus diseases is not accomplished by spraying or dusting, but by cultural or harvesting methods or by pruning. Some diseases attack only trees which are already weak while others attack those on stocks unsuited to the soil. In controlling diseases in general, good sanitary practices and maintenance of tree vigor by proper nutrition and soil moisture are factors of great importance.

Biological Control of Pests and the Spray Program

By no means all of the living organisms found on citrus trees are harmful; indeed, some of them play an important part in holding pests in check—a role spoken of as biological control of pests. Often they greatly decrease the need for sprays or dusts to be applied, and sometimes choice of a material for pest control should be made with due consideration of its possible effect on these helpful organisms.

Beneficial Insects—Predatory insects, which feed on other insects, contribute greatly to the control of some insect pests, sometimes giving a very high degree of control while being only partially effective in other cases. Thus the Vedalia ladybeetle is so effective in keeping down the population of cottony-cushion scale that spraying is almost never needed in bearing groves. The Chinese and blood-red ladybeetles feeding on aphids and the Australian (Crypt) mealybug ladybeetle feeding on mealybugs do not obviate all need for spraying although they decrease that need. These ladybeetles are rather specific in their choice of victims, but several other ladybeetles, and various lacewings, mealywings, and thrips, feed on a variety of insect and mite pests.

Parasitic insects, whose larvae devour the host from the inside after hatching from eggs laid in it, also help greatly in control of certain insects. There are several wasps which parasitize Florida red scale and small flies which parasitize aphids, but the outstanding examples are the Aphytis wasp which controls purple scale and a related species which attacks the Florida red scale. Certain insecticides have proved undesirable for use in spraying citrus trees because they were found to kill predators and parasites so seriously that there was more increase in insect pests following their use than if no spray had been applied.

The beneficial insects are today receiving increased attention, especially in areas where a regular spray program is not required and they may be a major reliance for control of insect pests. Furthermore, their activity often reduces or eliminates the use of chemicals which may have detrimental side effects on plants, animals, or humans.

Beneficial Mites—A few species of mites, which are to some extent predators of pests, are often found on citrus foliage. To a large extent they live on the remains of dead insects and mites, but some species are known to consume scale eggs and crawlers and others feed on six-spotted mites. The extent of their effective control is not well understood at present, but they certainly do not harm the tree and they are predacious to some extent.

Beneficial Fungi—Many kinds of fungi have long been known to hold in check various insect and mite pests, especially purple scale, mealybugs, whiteflies, aphids, and rust mites. Although some of the most conspicuous of these "friendly fungi," once thought to be of great value in pest control, have been found only to attack insects already dead, as is true of the redheaded scale fungus, others are of genuine effectiveness. The outstanding example is the very considerable degree of control of whiteflies by the red and yellow aschersonias, which make spraying for this pest rarely necessary so long as spraying for scales is done. The high rainfall of Florida and the resultant high humidity which are so conducive to development of diseases are also very favorable for growth of beneficial fungi.

Snails—For over a century the Manatee tree snail, considered native here, has been observed in Florida citrus groves. The largest populations are found in Lake County around Leesburg, where they have been increased by deliberate introduction of this snail to groves and by efforts to provide favorable environment, especially as regards humidity and winter protection. This snail is also found in some groves in Hardee, Sarasota, and Manatee counties, and perhaps others.

There is considerable evidence that the presence of sufficient numbers of this snail in citrus groves results in an increased glossiness of leaf and fruit rind and an apparent decrease in populations of insects and mites. In order to obtain beneficial effects from the snails, nearly all pesticidal sprays must be eliminated, but in some groves where spraying has been practically abandoned and snail populations have flourished, there has been no appreciable pest problem. Unfortunately, its true relationship with the total biotic environment is still imperfectly understood.

Physiological Sprays

Sprays are sometimes applied to citrus trees, not for control of pests but for affecting more or less directly the vegetative growth process or the rate of development of chemical compounds in the fruit. These are called physiological sprays since they have an influence on physiological functions of the tree and its fruit.

Nutritional Sprays—Sprays to supply mineral elements to the tree have been discussed fully in the chapter on fertilizers, since they really are just an alternative method of fertilizing. They could equally well have been discussed here, because very often the mineral salts are applied in the same spray mixture as insecticidal or fungicidal materials. Thus zinc is usually added to the postbloom spray made for control of melanose, and this same spray may also contain materials to control mites and scales if needed.

Maturity Sprays—Legal maturity of grapefruit is largely concerned with amount of solids and acid, with a minimum juice requirement. There is little change in the solids (sugar) content from August to

December, but acidity decreases rapidly during these months, and palatability is largely a matter of acid content being lowered until there is a pleasant blend of sweetness and acidity. However, there is great variation in the time when this ratio is obtained in different seasons, so that sometimes fruit is legally mature in November and sometimes not until April. Pink and red grapefruit varieties normally have lower acidity than the white varieties and so always have a passing ratio by December 15, and they sometimes reach this condition late in September.

Many years ago, in the course of using lead arsenate as a pesticide, it was noted the fruit on trees sprayed with it contained less acid than fruit on unsprayed trees. By use of arsenic spray it is possible to have grapefruit pass the maturity test in late September or early October, and all through the season fruit from sprayed trees will taste sweeter than fruit from unsprayed trees. The earlier it is desired to have fruit pass the test, the higher the concentration of lead arsenate in the spray must be and the earlier it must be applied. The longer the time for arsenic to act on the foliage, the lower its concentration need be to bring about the same reduction in acid. Incidentally, the effect is due to action of arsenic in the leaves, not in the fruit, and exceedingly little of it is in or on the harvested fruit. When arsenic is sprayed on orange trees the same decrease in acidity is obtained, but the resulting fruit is insipid and of poor quality. Florida law, therefore, prohibits use of arsenic on any citrus species except grapefruit.

To obtain fruit with earliest possible legal maturity a spray must be applied to grapefruit trees soon after the bloom period is over, before fruits reach 1½ inches in diameter, using 6.25 pounds of lead arsenate per 500 gallons for white varieties but only 3 pounds for red ones. This difference is because red grapefruit does not have minimum juice content early enough to take advantage of higher concentrations. This spray should be applied only in groves which have had an early bloom and may be combined with the postbloom melanose spray. It should assure maturity in October, or in occasional seasons in late September.

To assure that all fruit will pass the test in December and January, and also to have more palatable fruit in any month, the spray should have 2 pounds lead arsenate per 500 gallons for both white and red varieties. For sweeter fruit in February, March, and April, the spray need have only 1 pound of lead arsenate for all varieties. These applications for sweeter midseason and late season fruit may be put on at any time up through the summer period. Only a single application is needed, as a slow outside cover.

Excessive applications of arsenic should be avoided. Two applications a few weeks apart do not advance maturity more than the single one. High dosages for early maturity should not be repeated in the same grove year after year. It should always be remembered that arsenic treatments add to the cost of production, that continued use of high amounts can be detrimental to the growth and fruitfulness of the trees, that arsenic sprays make boron applications necessary, and that injudicious use of this technique for lowering acidity may reduce fruit quality.

Fruit-drop Sprays—Apple growers have long used sprays of certain chemicals to reduce dropping of fruits nearly ready to harvest, and for some time citrus growers have also been able to do this. Dropping of nearly or quite mature fruit—the preharvest drop—of 'Pineapple' and seedling oranges, and of 'Temples', can be greatly reduced by spraying with 20 ppm (acid equivalent) of 2,4-D. A single application should be made in October or November, as a slow, outside cover, and may be combined with a spray of wettable sulfur for pest control. The growth-regulator 2,4-D is used in higher concentrations as a weed-killer and will have very harmful effects on citrus foliage if great care is not used in preparing spray solutions. Enough of the material may remain in the empty spray tank to injure tender shoots when a subsequent pesticidal spray is made unless the tank is thoroughly cleaned, using washing soda (6 pounds per 500 gallons) to neutralize the 2,4-D.

Growth Regulators—As mentioned above, 2,4-D is a growth regulator in its physiological action, preventing preharvest drop of mature fruit. Considerable research has been, and is being, conducted with various chemicals which have influences on the physiological activities of citrus trees.

Spray applications of gibberellic acid (GA) applied between full bloom and petal fall have materially increased fruit set and yields of fruits of 'Orlando', 'Minneola', 'Robinson', and 'Osceola' tangelos. This material is approved in Florida for use on all tangelo varieties but is not recommended for other types of citrus.

Perhaps the greatest interest in growth regulations is in that group of chemicals which activates the development of the abscission layer and makes it possible to sever the mature fruits from the tree with less force. These abscission chemicals are being sought for use in conjunction with mechanical harvesting of citrus fruits. One of these, cycloheximide, which now has EPA clearance for early and midseason oranges, reduces the pull force by two-thirds and reduces, significantly, the percentage of plugged fruit, i.e., fruit of which a portion of the rind at the stem end remains on the tree. Since these materials are applied as sprays just prior to the harvest dates, production crews may be involved. Current recommendations should be obtained from experiment station or extension service personnel or from industry representatives.

10. Other Grove Maintenance Practices

THE programs of grove rehabilitation, fertilizing, and spraying, with their scientific foundations, have been discussed in the previous chapters. Four other programs are included in the total production schedule for citrus. Cultivation and cover crop management, and pruning should become standard practices according to their respective requirements. Drainage and irrigation, and cold protection, related to the soil and climatic conditions of the grove location, may involve considerable costs and extensive study. Expenditures for grove road maintenance, upkeep and repair of buildings and fences, etc., which do not fit into any one of the 7 standard programs, should be itemized as miscellaneous to obtain the total operational costs of the year.

Cultivation and Cover Crop Management

Soil management practices for the various tree-fruit crops have been developed primarily as means of creating or maintaining the most favorable soil conditions for tree growth, especially as related to soil moisture and soil nutrient supplies. They may also have a more direct influence on tree health through their effect on injury by pests or by cold. The systems of soil management in use in different parts of the country vary greatly with differences in soil type, in climate, in topography, and in the particular tree-fruit involved. In Florida, as in many other areas, the systems in use have been evolved chiefly by trial and error of growers, and, because they are largely independent of organized research, they often include practices which have no justifiable basis except that they have become customary.

Three systems of soil management have been used in Florida citrus groves—clean cultivation, continuous cover crops, and a combination of these two. The term "cover crop" properly refers to plants grown as a soil covering to prevent erosion by wind or rain. Various herbaceous crops, called "green manures," are also grown in orchards and fields for the organic matter they add to the soil when turned under before they reach maturity. In Florida the cover crop is grown mostly for the sake of its addition to soil organic matter, rather than for erosion control, and would more accurately be spoken of as a green manure crop in most instances. Since the usage is well established, it is not feasible to try to change it, but the situation should be clearly understood. The sandy soils of Florida are very low naturally in organic matter and benefit greatly from regular additions of it, so far as their ability to support good tree growth is concerned.

At one time clean cultivation all during the year was advocated by some growers, whose ideal was that no living plants should ever exist in citrus groves except the trees. This resulted in disappearance of practically all organic matter from the soil, so that its ability to hold water and nutrients became very low. The constant cultivation was expensive and the trees decreased rather than increased in vigor; consequently this system has been completely abandoned.

Other growers have attempted to keep the soil constantly covered by some herbaceous crops, planting summer annuals in the warm months and cold-resistant crops during the winter months. It is hard to get much growth of such green manures during the winter, however, because soil moisture is usually low. On the other hand, the cover crop plants compete seriously with the trees for the little moisture there is, and create hazards of fire and frost. For the sand hills this system is not feasible, although in groves on low, heavy soil there may be adequate moisture and the cover crop may be very helpful in preventing the breaking down of the edges of the mounds or beds on which the trees are grown. In southern Dade County the natural vegetation is allowed to grow all year long, because disking is impractical on the rocky soil, and the competition for moisture in winter is decreased by mowing.

The system most commonly used in Florida citrus groves is to grow cover crops during the summer rainy reason and practice clean cultivation during the dry period from early fall to late spring. Where this system is used in bedded groves, the cultivation employed is modified slightly to avoid injury to the edges of the beds. The heavy stand of cover crop which is encouraged to grow in the summer months does not compete with the trees, since the rainfall normally is ample for both, and does provide large amounts of organic matter (and of extra nitrogen, in the case of legumes) for improving the soil. As soon as the rainy season is past, the cover crop is turned under to conserve soil moisture for the trees, to eliminate fire hazard from dry plant material, and to prevent the additional possible cold injury due to interference by cover crop plants with radiation of heat from the ground and with flow of cold air.

Two general types of cover crops are used—leguminous and non-leguminous. Many growers favor legumes, planted in April usually, because of the additional nitrogen which they are able to take from the air and add to the soil through the agency of certain bacteria living in nodules on their roots. Beggarweed and cowpeas were popular at one time, then Alyce clover and the crotalarias were in favor, and more recently hairy indigo has been much planted. While it is quite true that legumes do increase soil nitrogen, the amount is negligibly small, so that no difference is made in calculating fertilizer needs in groves with legume cover crops from those with other kinds. Furthermore, these legumes constitute added expense for seed and seeding, or else require special timing of the autumn cultivation to assure that mature seed is disked in for next year's crop. And some of them, especially the crotalarias, create a problem in connection with plant bugs, as described in Chapter 9. Because of the rather small advantages and considerable disadvantages of legumes, many other growers prefer to make no special planting of a cover crop and simply allow the native or naturalized weeds to cover the ground. They feel that the amount of organic matter produced for a given expenditure is the more important criterion of cover crop worth. The cover will consist of various grasses, especially the pink-plumed Natal grass but even including sandspurs, and Spanish-needle and other broad-leaved weeds. A considerable amount of organic matter is grown at no cost, and so this system is probably most to be recommended.

As has been suggested, the cover crop may have a direct effect on tree health while it is standing, quite apart from its effect on the soil after it is turned under. This may be related to increased chance of cold injury on frost nights, as stated just above, or to certain insect pests, or to the ease with which a carelessly tossed match or cigarette may set fire to the dry, dead plants. Prevention of frost injury and fires has already been dealt with, but pest control by cultivation requires accurate timing and an understanding of how the insects develop.

Plant bugs and grasshoppers are the pests mostly involved. The former are a problem only in connection with a few legumes, on whose pods they feed by preference. If crotalaria, hairy indigo, or velvet beans are being grown in a grove, plant bugs will multiply on them, and when the cover crop dies or is disked down in late fall, the bugs move to the trees and feed on the maturing fruit. Such crops growing in groves or in adjacent fields should be examined in early September, and if heavily infested with plant bugs should be disked down at once while they are still mostly in the wingless stage. In cases of very heavy infestation an application of insecticide may also be desirable after disking.

Grasshoppers find favorable conditions in grass cover, especially crabgrass, either in citrus groves or in old neglected fields adjacent to groves, for multiplying to seriously harmful numbers. When this occurs they attack the trees as an additional source of food and may weaken them by partial or complete defoliation, sometimes leading to death of young trees. The principal grasshopper pest has two generations a year: (1) eggs laid in November and hatching in early May, with heavy feeding in May and June; (2) eggs laid in July, hatching in August, and causing injury mostly in August and September. Clean cultivation in the grove in November and December will prevent laying eggs for the first generation, while clean culture in April and early May will deprive any young 'hoppers of food. If grasshoppers invade the grove from fields in June, they must be controlled by chemicals to avoid immediate injury. Clean cultivation from August 15 through September will prevent any nymphs from the eggs of invaders from causing injury to leaves and fruit in the fall.

There will be only three or four months when the ground is left undisturbed for growth of the summer cover crop. To keep it clean of vegetation, or largely so, the rest of the year requires several cultivations. The following program has been fairly typical of what is common practice in citrus groves on fairly level, high ground. It is also commonly used in bedded groves on low ground except for special care not to break down the beds. About six cultivation operations are involved, whether harrowing, plowing, or mowing. This program is gradually being modified as herbicides become more generally used (see below).

1. *Late winter disking.* Immediately after the application of fertilizer in late January or February, the grove is cultivated with a disc harrow. While quite customary, this operation accomplishes no purpose unless unusually heavy winter rains have produced a volunteer cover crop which may compete a month or so later with the new flush of tree growth for a limited soil moisture supply.

2. *Late spring disking.* If a leguminous cover crop is to be planted, the ground must be prepared and the seed covered by use of the disc

harrow. If no cover crop is planted, the grove is harrowed right after the summer application of fertilizer. Unless there was a heavy infestation of grasshoppers the previous fall, so that hatching of many eggs is expected in May, this cultivation also is quite unnecessary and useless. There is no benefit from disking the fertilizer into the soil.

3. *Midsummer chopping or mowing.* Just before the summer application of a scalicide spray (late June or early July), the cover crop is reduced by use of a cover-crop chopper or a mower to allow greater ease of movement of the spray machine and the spray men, thus reducing the quantity of spray used. This operation also stimulates production of further green matter by the cover crop. A second mowing or chopping a month later may be worthwhile for the sake of more bulk of organic matter returned as green manure to the soil.

4. *Fireguard plowing.* In early September a fireguard is usually made with a moldboard or disc plow on those sides of the grove which are exposed to risk of fire hazard, since soon after this some of the cover crop may begin to mature and turn dry. The fireguard is preferably made just outside the grove, if possible, on the exposed sides, and in the first adjacent middles. Plowing in the middles should be shallow, to injure as few roots as possible, yet turn completely under all plants from tree row to tree row. The fireguard gives quick temporary protection until the whole grove can be disked.

5. *Early fall disking.* Late in September the grove is cultivated both ways (north and south, and east and west) with a disc harrow. If many acres of groves must be disked with the same harrow, the grower often cultivates all of them in one direction, and then starts the operation across the first disking to complete the job. In earlier grove practice the ground was all plowed at this time. While plowing has largely been superseded by disking, there is no good objection to a shallow (3- to 4-inch) plowing one way each fall except that it is more expensive. Unless very shallow, plowing may injure roots badly.

If plant bugs or grasshoppers are a problem, this cultivation will have to be done in late August or early September, as explained above. The grower must be aware of the amount and nature of cover-crop infestations, and govern his timing of cultivation accordingly. If insecticides must be used to protect the grove from invasion from adjacent fields, the disking should precede the spraying or dusting in order to save materials and application costs.

6. *Late fall disking.* The previous disking should have greatly reduced competition for moisture, but would not assure that the soil would remain free of vegetation all winter. Some cover-crop plants would persist and grow in October. So in November a final cultivation

of the grove in both directions is done to remove completely from the surface any vegetation which might increase frost injury or permit fire to flourish. If the grove was plowed one way in September, then it should now be disked first in the direction plowing was done, to level the soil, and then at right angles.

The discussion has been based hitherto on grove practices for fairly level ground. Where there is considerable slope, the citrus grower must modify cultivation, especially this last one, so that he does not make wind erosion easy. Erosion of this type in central Florida in groves made absolutely bare of cover has caused loss of much topsoil at times. Such erosion is largely prevented if there is some plant residue to anchor the sand. What is called a "trashy" cultivation is preferable to "clean" cultivation for erosion control.

Chemical Weed Control

It is always necessary to eliminate weeds around young trees (see Chapter 6), along fence rows, and around buildings. These operations formerly entailed the exhaustive and expensive manual labor of hoeing. It is now possible to use recommended herbicides with considerable cost savings and with better utilization of manpower. Furthermore, these chemicals may also be used for particular situations within the grove or nursery such as in and along ditches, in burrowing nematode buffer zones, and in spot eradication of a serious weed encroachment offering competition to the tree. Milkweed vine and balsam apple vine may be reduced in the grove area by a combination program of manual and mechanical operations with properly timed herbicidal applications.

The milkweed, or strangler, vine, *Morrenia odorata,* an import from South America, was cultivated in Pasco County in 1939 and was found infesting groves in central Florida in 1960. By 1970 it had spread into groves in Lake, Orange, and Seminole counties and as far north as Marion County and south to Highlands County including the coastal counties. It grows and reproduces rapidly, shading out the trees and even girdling limbs by growing around them for support. It will take a number of years of mechanical and chemical treatments to bring heavy infestations of this weed under sufficient control to allow a maintenance program to be instituted. Balsam apple vine, *Momordica charantia,* is a serious problem in coastal areas and is found in other parts of the state. Perennial grasses such as bermuda, guinea, bahia, torpedo, and vasey, known in Florida for many years, have often become serious weed problems in Florida citrus groves. These can be controlled with recommended herbicides.

Research has developed many new herbicides which act systemically, by contact, or by soil sterilization. These will gradually supersede mechanical weed control with reductions in costs and labor requirements. All of them require approval by EPA prior to commercial use. Care must be taken to see that undue amounts on citrus foliage, or excess amounts in the soil, do not bring on toxicity symptoms in the trees. Because of the diversity of their uses, in ditches, nurseries, young groves, bearing groves, spot treatments, and barriers; their varying dosages for different soil types; their timing and soil treatment requirements; and, especially, because of rapidly developing new products and techniques, the grower should avail himself of the latest information from extension and commercial personnel in the field.

Irrigation

Irrigation is not a necessary operation in Florida citrus groves on deep, well-drained soils; many groves have survived productively with reliance on rainfall even though periods of water stress within the trees have occurred. Nevertheless, with sound management, the program has been profitable. In groves on shallow soil, in which drainage facilities are required during rainy months, irrigation practices become an obligation during droughts. To understand why this is so and what factors determine the desirability of irrigation, we shall have to consider the needs of plants for water and how those needs are satisfied.

Water Needs

Protoplasm, the living substance in plant (and animal) cells, is able to carry out its functions only when it contains a good deal of water. In a dry seed, the protoplasm is alive but almost completely inactive. The first step in germination is absorption of a large amount of water, until the liquefied protoplasm is able to carry on metabolic activities, divide, and grow. Most plant tissues die long before the protoplasm of their cells ever reaches the low moisture content of a dormant seed. The need for water by protoplasm is one of the basic realities of plant life.

Water has many uses in plant tissues besides making protoplasm active. No mineral nutrient or gas can enter plant cells unless dissolved in water, and no organic compound can move from one cell to another except in solution. It is the universal solvent, and the medium in which all chemical reactions occur. Water is itself a nutrient, as one of the raw materials in the process of photosynthesis by which sugars are manufactured in green cells with light energy. It is in water as a vehicle that mineral salts move from root tips up through trees to the leaves. Tender new shoots and young leaves are enabled to hold themselves erect or spread out horizontally without any rigid framework because their cells are distended or turgid with water. Only when plant cells are turgid can they grow larger or divide, and only when leaves are fully spread out can they make effective use of light. In times of moisture shortage this turgidity is lost and leaves collapse into a wilted condition.

The amount of water needed for all these uses is really very small, however. The big problem of land plants is that they constantly lose water from their tissues in the process called transpiration. The amount of water required for all other purposes is infinitesimal compared with the amount transpired daily from leaves, stems, and fruits of citrus trees. So long as the atmosphere surrounding the plant is not saturated with water vapor, the plant must lose water, and the drier the air, the faster the rate of water loss. Unless this water can be replaced promptly, the plant suffers, and so the possibilities of replacement are of vital concern. Practically all the water used by trees is absorbed by their roots from the soil, and so we must examine the ability of soils to supply water.

Soil-Water Relations

In the discussion of soil in Chapter 4, it was pointed out that the amount of available water which a soil could hold was determined by the difference between percentages present at field capacity and at the wilting point.

A typical Astatula fine sand at the permanent wilting point has about 1 per cent of water, at field capacity about 5 per cent, and is able to hold (by difference) a maximum of about 4 per cent of available water. A soil of the Arredondo series, a phosphatic soil with slightly more organic matter, holds about 2 per cent at the wilting point, 8 per cent at field capacity, and so has around 6 per cent of available water at maximum capillary capacity. Thus each soil has its own characteristic water-holding capacity.

It would appear at first that the soil with the highest percentage of available water would be the one which could supply the most water to a citrus tree. But the total amount of water available to the tree is dependent also upon the volume of soil permeated by its roots, and this means primarily the depth of the soil to a water table or hardpan. A deep, well-drained soil with an effective rooting depth of 5 feet and a maximum of 4 per cent of available water can hold for use by trees nearly twice as much water as a soil with 7 per cent of available water at field capacity but only 1½ feet deep to a hardpan. That is why low, moist soils often have greater need of irrigation than deep, sandy soils. During a drought the shallow soil must be irrigated more frequently, and can absorb less water at each irrigation than is the case with a deep soil.

Absorption of water by plants is not the only way, however, in which available water is lost from the soil. Evaporation from the soil surface removes much water, especially from the top foot of soil, for there is little loss by evaporation below this depth. More of the total available water is lost, therefore, from a shallow soil by evaporation than from a deep soil. This is a further reason why shallow soils must be irrigated more frequently during dry periods than deep soils.

Plant-Water Relations

The leaves of a tree with plenty of moisture in it are turgid, but as the water content of the tree is reduced to a condition of moisture deficiency, the leaves gradually lose turgidity and become wilted. The first indications of this condition, called incipient wilting, are found in early afternoon when transpiration exceeds absorption of water. This incipient wilting is not visible but can be demonstrated by suitable methods. Such a slight moisture deficit in the leaves occurs almost daily, except in rainy weather, and seems to have no harmful effect at all.

As the moisture deficiency within the tree becomes greater, however, as the result of a succession of days when transpiration exceeds intake, the leaves begin to show a visible temporary wilting for a short time each afternoon, with the length of time when this is evident increasing daily. This temporary wilting disappears later in the afternoon at first, as decrease in transpiration enables the roots to replenish the moisture shortage, and so long as it lasts only an hour or two there is no great loss of food-making resulting from it. The importance of temporary wilting is its indication that the soil moisture is approaching the permanent wilting point and irrigation may be needed soon. As the moisture shortage continues, the leaves wilt earlier and recover later, often not until after dark, until eventually the time comes when they fail to recover turgidity at night. The soil has now reached the permanent wilting point and the leaves will remain wilted permanently unless water is added to the soil. If the soil water supply is not increased within a certain length of time, the twigs and limbs dry out and the tree may die; or at best it will lose all its leaves. If water is added by rain or irrigation soon enough, the wilted leaves can recover and the tree will have suffered no great permanent injury, although more or less dropping of leaves and fruits may have occurred.

Irrigation is most beneficial if supplied as soon as temporary wilting is observed. Some growers also watch cover crops because these plants have a much shallower rooting zone than the trees and tend to show moisture shortage by wilting before the trees do. It might well be considered that it would be even better to make physical measurements of the soil moisture and be aware all along of the growing shortage of moisture, instead of waiting for plants to show it. Methods and instruments are available for doing this, but unfortunately the amount of sampling which must be done in a citrus grove to obtain reliable values for soil moisture make measuring too expensive even for the large, commercial grower, let alone for the small grove owner. An accounting system, based on daily losses of water by evaporation and transpiration (evapotranspiration, which accounts for the greatest loss of soil moisture) and additions of water through rainfall and irrigation, is being used effectively to determine the needs of supplemental water. Such a system should be based on the type and depth of the soil in the individual grove; it requires diligent efforts on the part of management but does take much guesswork out of the program.

Drainage

It may seem paradoxical that drainage should be a factor in irrigation, but for the citrus grove on low ground this is often the case, and more than 50 per cent of the citrus groves in Florida are now planted on poorly drained soil. The early citrus growers often considered the high, well-drained locations satisfactory as to drainage but very subject to drought, while the low, poorly drained lands were recognized as presenting drainage problems but considered as having little need for irrigation. Today it is generally accepted that the deeper the soil, the larger the rooting system which a tree can develop and the less the need for irrigation. The poorly drained soils present the biggest irrigation problems.

Need of Irrigation

About once in every ten years Florida experiences a drought of serious economic consequences; about one year in three the spring rainfall pattern does not provide sufficient soil moisture to take care of the needs of citrus trees properly. The critical period is usually during the season from February to April, when fruit is setting and starting to increase in size. Most growers who expect to irrigate keep close watch on trees and weather records during this period. As soon as the rain-

fall record shows that rainfall is below normal for the season, the grower should be alert to detect the first temporary wilting, indicating that irrigation should be begun. Too often, however, the grower waits in hopes that rain will come and make irrigation unnecessary, until the leaves are already badly wilted before irrigation is started.

In most sections of Florida, water has been readily available from lakes or shallow wells, and distribution through permanent or portable mains easily done. Power sources and labor supplies have made possible extensive irrigation. Unfortunately, analyses of the economic results from irrigating have not shown that it was always a profitable operation. Growers have applied water when it was not really necessary, or have applied it too late to obtain benefit. Often a heavy investment in irrigation equipment has been made without any clear ideas on what use was going to be made of it; such equipment was just something a grove should have. And even when the equipment is bought with clear understanding of what could be done with it, it is easy for the grower to spend more money irrigating than he can gain from the operation unless he analyzes carefully the factors involved.

Fall irrigation may be worthwhile occasionally if the trees show severely wilted foliage or fruit begins to wilt and become flaccid before ready to harvest, but such occasions will be very infrequent. As a rule, fall irrigation is undesirable because it may induce a flush of growth which will be more easily injured by winter cold and will leave fewer buds to produce blooms in spring.

Irrigation in the spring, when temporary wilting appears, in a season of less than normal rainfall will increase the yield of fruit and is usually an economically justified operation. Even then, irrigation should be as limited as seems compatible with tree health and fruit retention, for unnecessary application of water not only adds to the cost of production but also tends to cause development of fruit with lower content of solids, especially when applied late in the season.

In most Florida soils used for citrus groves an acre-inch of water will wet the soil to a depth of about a foot. Deep, well-drained soils with an effective rooting depth of at least 4 feet should have about 2½ inches of water at each irrigation. Shallow soils must have less water applied, depending on their rooting depth. For example, in a bedded grove with only 18 inches in which roots can develop, not more than 1 inch of water should be given per irrigation. Obviously several successive applications will be needed here to provide the same amount of water to the trees which was given in one application on the deep soil, and on such shallow soils more of the applied water will be lost by evaporation.

Irrigation Equipment

This program should be developed with the proper understanding of the water availabilities; labor, power, and equipment; and the basic operating conditions. An economic feasibility study should be made. The early citrus grower employed various means of water distribution requiring high labor inputs, and often simply used the furrows between rows of trees in bedded groves. The gasketed, lockjoint, aluminum portable pipe system came into vogue and was highly satisfactory during the 1940's. Surface and deep-well water sources were generally available; gasoline and electricity were the power sources; centrifugal and turbine pumps presented no problems. Currently, with large expanses of groves, water availability has become more of a problem and labor for the portable irrigation line almost nonexistent. Systems which require higher capitalization but far less labor had to be developed. Volume guns, which water large areas of one-half acre or more, with underground feeder lines; self-propelled mobile guns which travel down the middle and require manpower only when being moved from one middle to another; and, finally, the solid-set system of permanent risers with butterfly sprinklers set along each tree row to cover the entire ground area, have come out of sheer necessity. The latest interest is in the drip, or trickle, system of distribution of water through shallowly placed PVC lines and discharge through emitters (drippers) under each tree. Many advantages and disadvantages can be found for each method of distribution which should be considered in the feasibility study.

Pruning

Pruning, the cutting off of parts of a plant in order to develop a desired shape or to remove dead, diseased, or poorly placed branches, has been practiced with many kinds of fruit trees for many centuries. The early leaders of the Florida citrus industry were mostly men with previous experience in growing deciduous fruits in the North, and they tried at first to use the same types of pruning for oranges which they had learned to use on apple trees. Gradually they realized that there is no need for training to a particular system, since citrus trees have naturally strong crotches and limbs seldom split away, and that while the trees grew well with different types of pruning, they were apparently about as thrifty and productive with a minimum of pruning as with a good deal. Since pruning adds to the cost of production, growers have come to abandon pruning for form and vigor, and pruning of bearing trees is largely for reasons of health or size.

Pruning needed during the first four years of the citrus tree has been discussed in Chapter 6. In the early, bearing years, from the fifth to the tenth year, little pruning is required except removal of water sprouts as explained below. For trees older than ten years, more pruning will be needed for various reasons. The following pruning practices are fairly standard in mature, bearing groves.

Sprouting

Once a year, preferably in the months of April or May prior to the summer application of spray for scale control, water sprouts should be removed from the trunk and scaffold branches of oranges, tangerines, and other citrus species which tend to have many competing branches arise in the main framework area. Failure to remove such sprouts regularly reduces the fruit-producing ability of the tree by diverting water and nutrients to unproductive shoots, and increases the problem of control of whiteflies and other pests which feed on these succulent tissues. Control of these pests is more costly and uncertain if done before instead of after this sprouting operation. All shoots should be pruned which are so placed that they cannot become desirable permanent branches, but fruiting twigs, which are usually round in cross section instead of angular and up to 7 or 9 inches long, should not be removed from the interior of the tree; yield would be thereby reduced unnecessarily. Any shoots which will fill in a vacant space in the periphery of the tree should be left to become part of the permanent framework. All these sprouts should be removed when so small that they can easily be cut with hand clippers. If older sprouts are present, overlooked in past years, they may require use of long-handled pruning shears. The work can be done rapidly but calls for exercise of some judgment.

Deadwood Pruning

There are various reasons for the presence of dead twigs and branches in a citrus tree, among them waterlogging of soil, infestations of insects and mites, and attack by fungus diseases. Simply removing such deadwood without determining and correcting the cause is poor grove management. Hurricanes, freezes, and droughts, which are largely beyond man's control, may cause dying of large limbs as well as small. Even without these external causes, twigs regularly die because growth of other branches shuts them off from light until they can no longer make enough food to maintain themselves, and others may be so weakened by the flow of food from them to a developing crop of fruit that they die of starvation.

Twigs which die from being too shaded will self-prune themselves and might be left to do so, and much other deadwood might be left to erode away except for its unattractive appearance and its causing scars by rubbing against developing fruit. However, dead twigs are an important source of melanose infections on fruits, and while it is not feasible to do a sufficiently thorough job of pruning of fine twigs to control melanose by pruning alone, removal of these infected twigs is very helpful as a supplement to spraying. There is also the fact that as larger limbs die back following freezes or fungus attacks on the roots, they may carry decay back into permanent trunk or main limbs. Every three to five years, therefore, a pruning crew should go through the grove with lopping shears and pruning saws to cut out the dead branches of a diameter of ¼ inch or larger. It is not economically feasible in commercial practice to try to prune out all small dead twigs, however desirable this would be in theory. Usually from 30 minutes to an hour per tree will suffice for this pruning. Lopping shears will easily cut branches to ¾ inch in diameter, but larger limbs should be cut with a saw. All cuts should be made flush with the larger branch from which the dead branch arises, or with a lateral branch in heading back limbs, so that no projecting stub remains; and the final cut should always be below any dead tissue. It is usual to treat any cut surface over an inch in diameter with pruning paint although even larger cuts may heal well without such protection when made on limbs with many healthy leaves supplying food for healing. This pruning operation is best done during the summer months because labor is most available for such work then. Furthermore, by this time it is possible to tell just how much damage was done if there was a freeze during the winter, so that cold-injured wood can be pruned with assurance. It is always a mistake to try to prune shortly after a freeze.

Special Pruning

There are several types of pruning which are sometimes needed and can be done along with pruning of deadwood.

Long, thin branches which bear little or no fruit because they are not well enough exposed to light, and so will never be able to bear fruit any better in the future, should be removed so that their supply of water and nutrients will be available to better placed branches.

Tangerine trees may profitably be opened up by pruning out portions of the dense centers to allow more sunlight to penetrate the interior (making fruit color better) and to reduce slightly the number of fruiting branches (giving better fruit size). This objective is better accomplished by bulk pruning of a few large limbs than by pruning out many small branches. If the latter is done, it will be only a short time before the interior of the tree is as dense as ever. The earlier in the season of fruit development that this pruning can be done, the better.

Grapefruit trees, because of the great weight of their fruit, are especially likely to have some of the lower limbs bending down until they are nearly on the ground. The fruit on these low-hanging branches is often of poor quality and is likely to have cuts or other injuries made by cultivating equipment. These branches also complicate the problems of pest control. Certain insects, principally scales, accumulate under the soil which tends to coat the bark, leaves, and fruit, and are very hard to reach with sprays. They then form centers of rapid reinfestation of the rest of the tree.

Raising the bottom of the canopy of these trees, called "lifting" or "lifting the skirts," is easily accomplished by pruning off those limbs which are lying on the ground or very close to it. It is usually best to remove an entire branch at its origin on the trunk. To avoid having it tear away some wood and bark from the trunk, the limb should first be cut upward for at least one-third of its thickness. In the case of quite large limbs, this first cut is made several inches out from the trunk and the downward cut made half an inch farther out. After the bulk of the limb is gone, the stub can easily be cut next to the trunk. In either case the final cut should be flush with the trunk and the surface painted for protection. If removal of a whole branch would leave quite a prominent gap in the canopy of the tree, then it may be better to prune only those subordinate branches which hang near the ground. The best "lifting" is accomplished when the remaining branches are pulled down when loaded with fruit until the fruit barely or not quite touches the ground, and when relieved of the weight of the crop these branches do not hang lower than a foot from the ground. Raising the bottom of the canopy too high will reduce the yield of the tree, but a judicious "lifting" will involve loss of only a small amount of low-grade fruit which is more than compensated for by fruit of better quality in the higher portions of the tree.

A method of "lifting" which has recently been introduced in some groves is to use a powered circular saw. This is passed under the tree at a height of a foot or so and cuts all branches which hang lower than this. The pruning is quickly done, but such a saw is dangerous to handle and should be operated only under careful supervision.

Hedging

Since 1950 the process of "hedging" citrus trees has been developed in the state. Primarily it was developed as a remedy for the situation created in many groves where trees had been planted in rows 20 to

Fig. 10. A "hedging" operation in Polk County.
Courtesy Lake Garfield Nurseries Company

25 feet apart and had grown so large that they were badly crowded. By means of a gang of circular saws the rows are hedged, with the sides vertical or slanting toward the center of the tree at a small angle, so as to leave a clear middle 7 to 9 feet wide between tree rows. This admits light to the lower part of the canopy, increasing the bearing area and improving the color of fruit on the lower branches, as well as reducing costs of spraying and picking.

Any one of several systems of hedging can be used, depending on

how many sides of each tree need trimming and the frequency of the operation. One of the most satisfactory systems is to start the hedging operation at the time when the trees first begin to meet in the operating middle. By opening every other middle to good operating width (i.e., to about 8 feet) at this time, and after two years opening the alternate middles similarly and shifting operations to them, there is provision for continuous good operation. In the fourth year the first set of middles can again be opened, and the rotation followed regularly. In this system hedging is done in one direction only, which is the direction of most convenient operation. Another possible system would be to open the middles in one direction at first, and then a few years later to open the middles at right angles.

Hedging should be done in the spring, after the last of the old crop has been harvested and before the fruit of the new crop has made much growth. It can be done with a gang of circular saws in line or on rotating arms, driven by power take-off or an independent engine, and can also be done with hand pruning equipment by men on platform trucks. Similar equipment can also be used to cut a flat top across the tree at a desired height. There is some loss of fruit the year following hedging, but over a period of years the yields will be as great and fruit quality will be higher. With the advent of closer spacing in planting groves during the past decade, hedging, topping, and lifting are being given definite consideration as anticipated practices in the long-term projection of each individual grove. The soundness of these practices has been proved by the maintenance of maximum fruit-bearing volume of trees per acre by mechanical means, the reduction of insect and mite pests and reduced spray costs, the reduction of equipment damages and operator injuries in other grove operations, and production of increased percentages of marketable fruit.

Pruning Very Weak Trees

Following severe attacks of soil fungi or injury by hurricanes or freezes, old trees may have a tremendous amount of deadwood in proportion to that which is still productive. Unless vigorous growth gives evidence of prompt recuperation, the grower may well consider whether this cost of heavy pruning might not better be diverted to removal and replacement with young trees. Tree surgery, including filling of cavities and "wiring" of vigorous shoots from severely headed scaffold limbs, is hardly profitable in commercial groves. Trees in need of extensive and time-consuming repairs are better replaced, unless they are dooryard trees with sentimental value to the owner. (See Grove Rehabilitation in Chapter 7.)

PRUNING TECHNIQUES

As pointed out under "Deadwood Pruning," it is important in any kind of pruning to cut back to sound, live wood and to avoid leaving any projecting stub. Development of the healing callus can take place only from living cells and is dependent on a supply of food from the leaves. If the cut surface is in the line of flow of food down the branch, callus will develop satisfactorily, but if it is offset from the flow an inch, at the end of a short stub, no food reaches it and healing is impossible. If the cut is made flush with the larger branch, or is made at the point where a lateral branch arises, then the wound callus will be in the line of food flow. Speed of callus formation also depends on the rate of cell division, and this is much higher in spring than at any other season. The best time to prune, so far as healing of wounds is concerned, is just before the spring flush of growth, and the poorest time is in the fall and early winter; no pruning should be done from November to January.

If, for any reason, large areas of trunk or limbs are exposed to the sun as the result of heavy pruning or of defoliation by cold or storm, the exposed surfaces should be protected by a coat of whitewash or wrapped with burlap to prevent sunscald. In spite of their thick bark, these surfaces have been shaded by foliage and cannot endure the sudden exposure to direct sun. Sunscald can occur in winter as well as summer, but whenever it takes place, the trunk may be injured so seriously that the tree must be removed. Severe heading back to the top so that all foliage is removed is best done in spring when new shoots can be expected to renew the shade most quickly, with temporary protection of the bark meanwhile as above suggested.

To paint or not to paint pruning cuts is not nearly so important a decision for the grower to make as whether to make clean cuts which the tree can heal readily, with no projecting stubs. And speaking of projecting stubs, each hedging operation, especially when the operation has been overly delayed, should be followed by pruners who will cut off the stubs left by mechanical equipment. If not removed, these stubs will shortly be hidden by new growth and will present a danger to operators and equipment in future work in the grove.

Protection from Cold

The point has frequently been made in earlier chapters that injury by low winter temperatures is the greatest factor limiting the area for production of citrus fruits in Florida. A full discussion of the nature of cold-hazard in the various citrus-growing areas of the state will be found in Chapter 4, and in Chapter 5 there is a consideration of the selection of sites for groves with greatest natural freedom from cold-injury and of stocks and scions with cold-hardiness.

Having made the best available selection of site and chosen a combination of stock and scion suited to the general area, the grower can further avoid injury from cold by certain cultural practices. Banking of young trees has been discussed in Chapter 5. Trees which are free from nutritional deficiencies and from weakening by pests will endure cold better than trees that are weak for any reason. In the section on *Cultivation* in this chapter, the effect of a cover crop on cold injury and the importance of clear culture in winter are made clear. Finally, the grower can avoid practices, such as heavy application of nitrogen in late summer and maintaining high soil moisture in the fall and winter, which tend to maintain trees in active growth in the fall. Such trees are more easily injured by cold than trees which have early become dormant.

Finally, when the grower has done all he can by careful planning and good culture to help his trees escape injury from low temperature, there remains only for him to take steps to prevent temperatures from reaching such low points that trees will be injured in spite of their condition. The means most commonly employed are by heating the air within the grove, usually called "firing."

Firing

Until the "big freeze" of 1894–95, citrus growers in Florida seem to have relied on favorable location for protection from cold. The first grove-heating was done during the freeze of 1899, although far more growers then were in favor of covering trees with tents or lath (perhaps with supplementary heat) than of heating trees in open air. The cumbersome nature of covers and the good effects obtained with wood fires led to extensive grove-heating after 1900. Since pine wood was abundant, resinous heartwood of slash pine (called "lighter wood" or "fat pine") was commonly employed for many years.

The last pine wood supplies were used up many years ago; growers turned to No. 2 diesel fuel as a source of heat. At one time the black smoke produced by incomplete combustion was considered desirable and beneficial; smudge pots, which produced copious amounts of soot and smoke, were believed ideal for this purpose. It is now well recognized that it is heat, radiant and convective, not smoke, which protects trees from cold. Stack heaters of various designs and pressurized oil systems, assuring more complete combustion, have gradually replaced the smudge pots. Central systems are popular because they are clean, more efficient, easy to ignite, and require no labor for refueling. Furthermore, with increased urbanization the citrus grower is definitely restricted in the range of equipment which allows proper pollution control.

Even these more efficient sources of heat do not always assure freedom from cold injury. Sometimes the grower waits too long, until some freezing has occurred, before lighting the heaters, and then has no benefit from the money spent for heating. Sometimes cold winds lower the temperature so far, and carry away the heated air so rapidly, that it is impossible for the heating to give protection; again the cost of heating may be wasted. Sometimes the cold period continues for two or three days or possibly longer, until the supplies of fuel oil have been exhausted, and though the grower kept his trees from freezing for two days, they were injured on the third. But when the conditions are not too unfavorable, heating is very effective in keeping grove temperatures above freezing.

Thermometers are necessary, so that the grower knows how well he is controlling temperature. There should be at least one thermometer within the grove and another outside the grove but close by; the former enables the grower to keep track of the temperature around his trees, while the other tells him whether the surrounding air mass is getting colder or warmer. These thermometers must be sheltered from free radiation to the sky, so that they record air temperature, and from direct exposure to wind, and by standard usage are placed horizontally at a height of 4½ feet above the ground. An error of 1°F. in the reading of these thermometers can have serious consequences in fruit or twig injury, so they should be checked for accuracy in the critical range (20 to 35° F.) each fall before the first need to use them. The Federal-State Agricultural Weather Service, Lakeland, conducts tests of thermometers each fall in cooperation with personnel of the Florida Cooperative Extension Service. See your County Extension Director for details of this service.

Usually fires are lighted when the thermometers reach 30°F., and they should be kept burning until the temperature outside the heated area rises above the freezing point (32°) again. With either a light breeze or a cover of clouds, temperatures do not fall as low as on calm, clear nights, and these are the ones on which heating is most often needed and effective. Heaters should be spaced more closely on the side or sides of the grove from which drift of cold air is expected, so that the air is warmed as it enters the grove and maintains safe temperatures as it moves through the grove with much less assistance from other heaters. Otherwise 50 to 60 heaters per acre are commonly used in the portions of the grove needing heating.

However, even with favorable conditions it is considered that a differential of about 8° F. is the maximum which can be maintained by heating between the temperatures inside and outside a grove when the outside temperature is 32° or below, and in practice the difference is rarely this great. Growers generally tend to overestimate the amount of heat which is produced and are surprised to find later that there was some tree injury in spite of heating. On nights of only moderate cold, the grower might be well advised to estimate the damage likely to occur if he does not fire, and compare this value with the cost of heating, for heating is an expensive operation. At any rate, the grower should avoid making overoptimistic estimates.

Other Methods

Sprinklers are sometimes used, especially to protect nurseries or specimen trees. Even though ice may form on twigs, the internal temperature of the plant tissues cannot fall below 32° F. so long as the sprinklers are running. Ideally they should envelop the trees in a fine misty spray, but it is usually impossible to regulate the sprinkler heads of standard irrigation equipment to do this. Once started, the water must be kept running until the air temperature gets up to 32° and ice begins to melt. If the temperature falls far below freezing, there may be a thick coat of ice formed and its weight may cause much breakage of limbs. Nursery trees are so low that the ice usually forms a framework resting on the ground and does not put any load on the branches. Sprinkling is feasible only for groves so small that the sprinkler system will cover all the trees.

With the development of permanent irrigation systems with risers spaced regularly through groves, it seemed logical to adapt the systems for cold protection during the winter. This principle of water use was employed during the cold wave of December 1962, with disastrous results. Applying the water at the rate of 0.1 inch per hour allowed it to freeze immediately, subjecting the wood and foliage of the trees to lower temperatures than would have been experienced with no protection. Severe damages, with excessive cold cankers, occurred while adjacent trees, left unsprinkled, often showed no damages. Aside from the problems of keeping mud daubers (wasps) from building nests in the sprinkler mechanisms and the actual freezing up of the sprinklers, there is the necessity of applying sufficient water to maintain a slush ice on the trees at all times during the entire freezing periods. For lack of good information, and fear of a repetition of the events of 1962, overhead sprinkling as a method of cold protection in bearing groves is today discouraged.

Wind machines are also used in Florida groves for cold protection. Like oil heaters, they were used for some years in California before being tried here. The machine is essentially large single or double airplane propellers, mounted on a tower so they can revolve in a circle and operated by electricity, gasoline, or diesel oil. The principle on which they are based is that on still, clear nights there is a stratification of air, coldest next to the ground and increasingly less cold up to a height of anywhere from 40 to 200 feet, above which temperature falls again with height. This height at which this change occurs is called the inversion point, and the increase in temperature with height up to this point is termed a temperature inversion, since normally air temperatures are colder with altitude. If the inversion is large, that is, if the air temperature 40 feet above the ground is several degrees warmer than the temperature near the ground, then a wind machine can blow this warmer air down and mix it with the cold air. If there is a breeze, or if the night is quite cloudy, there is likely to be little or no inversion. The use of these machines must be tailored to the particular grove, and no recommendations can be made without study of the individual grove.

Cold Injury to Fruit

All the methods of cold-protection aim to keep grove temperatures at 30° F. or higher, though it is wasteful of fuel to keep them above 32°. Citrus fruit is first damaged when the temperature has fallen to 28° and remained there from two to four hours. The first evidence of injury is the presence of ice crystals in the fruit early in the morning following the low temperature. When the ice has melted, tiny white crystals of hesperidin are formed along the walls of the fruit segments, especially in sweet oranges. These crystals do not indicate that the fruit is harmful to eat, but do indicate in most cases that the fruit has been frozen, although in rare cases they may be formed from other causes. If the freezing has ruptured the juice sacs near the stem end, fermentation is likely to start in a few days, and then several days later the tissues at the stem end begin to dry out, with twisted segment walls. While fermentation is going on it is better not to eat much fruit, but either before or after this the frozen fruit is perfectly healthful, although containing less juice when some of the juice sacs are dry.

Fruit which has been damaged by freezing may show a water-soaked appearance or evidence of previous water-soaking, ruptured

LOOK FOR DRYING IN FRESH FRUIT.

juice sacs, mushy condition of the pulp, or dryness, depending on the severity of freezing and the length of time since freezing occurred. These damages are almost always at the stem end of the fruit and only rarely found at the stylar end.

Two types of freezing damage to fruits are recognized by the citrus laws of the state of Florida—*damaged* and *seriously damaged. Damaged* fruits are those showing drying or other evidence of freezing at the stem end for over ¼-inch depth in oranges or grapefruit, or ⅛ inch in mandarins, or an equivalent fruit volume, if occurring elsewhere than the stem end. If the extent of the freezing injury is more than twice the minimum for *damaged* fruit, it is considered *seriously damaged.*

Another, slightly different, form of cold-injury is produced in small citrus fruits by freezing due to direct radiation of heat to the sky. When any fruit is fully exposed to the sky, it radiates heat from its upper surface out into space. If this loss of heat continues until the fruit temperature falls below the critical point, the cells in this area are killed. Decay organisms invade the injured rind and rot develops where the cells are dead. This type of injury is rarely seen in oranges, grapefruit, and tangerines, probably because they have very thick rinds, but it is often found in the small-fruited citrus species such as calamondin, limequats, and kumquats, which are much grown in cold areas.

The Florida Department of Citrus, in cooperation with the Citrus and Vegetable Inspection Division of the State Department of Agriculture, maintains constant field service during and following any freeze to determine the extent of damages and to make sure that only fruit of sound condition is sent to market. Shipping holidays are also imposed following a severe freeze so that cold-injury will have time to be manifest before fruit is shipped; when shipping is resumed, fruit inspectors examine carefully all fruit packed for shipment to eliminate any showing freeze damage.

Severe freezes kill many twigs or even large branches. Growers often hasten to prune out this dead wood, but experience has shown that such pruning should not be attempted until after the spring flush of growth has matured. If many leaves have been lost and much deadwood produced, an application of neutral copper spray containing 3.75 pounds of metallic copper in 500 gallons may be advisable as the new leaves are expanding to protect them from heavy melanose damage. Addition of 25 pounds of wettable sulfur per 500 gallons will increase the effectiveness of the spray. Tender young leaves are hard to wet and sprays spread poorly over them; the sulfur acts as a wetting agent and spreader. Trees injured seriously by cold need to make unusually vigorous vegetative growth and so will need more fertilizer and water than trees of the same size that have not been subjected to cold damages.

Costs and Returns

Frosts and freezes decrease profits, for groves which have been injured by them cannot be operated at greatest efficiency. In addition to loss of fruit immediately, there may be decreased crops the next year or next several years. If crop and trees are saved by some method of cold protection, there has been a great increase in cost of production of each box of fruit and thus a decrease in expected net returns, although sometimes the loss of a large portion of the crop raises the value of what is uninjured enough to compensate for the cost of protection. The prospective purchaser of a grove needs to be especially careful to inquire into the history of cold-damage. On a warm summer day or during a mild winter, when there has not been a freeze for two or three years, it is easy to be overoptimistic regarding freezes. If one has not seen the grove following a period of freezing weather over the state, it would be well to seek advice from someone who knows well the freeze history of the area.

11. The Crop—Harvesting, Maturity, and Grade

THE PURPOSE of a citrus grove is, of course, to produce a crop of fruit, and if the procedures hitherto discussed have been faithfully carried out, there should be a good crop to harvest every year unless climatic catastrophes have intervened. Two questions which arise are when to harvest this crop and how to do it. The latter question is more easily answered and so will be taken up first.

If the grower belongs to a cooperative marketing organization or sells his fruit on the tree, his fruit will be picked by packinghouse crews and he will have no problem of picking method. But, if he has his own pickers he should be very conscious of the fact that a well-grown crop of fruit can be greatly reduced in value by carelessness on the part of pickers and handlers. Fruit which is going to the processing plant is utilized so soon after harvesting that picking care is not of great importance, but fruit destined for the retail market has plenty of chance to break down before it reaches the consumer if care is not used in harvesting. In spite of its leathery rind, citrus fruit cannot be handled roughly and stay in good condition, and any cuts or punctures made in picking offer easy entrance for decay organisms.

Prior to 1900, most citrus fruit was usually pulled from the twigs, often carelessly, and sent to market after a few days of curing in the packinghouse. A great deal of the fruit was injured by "plugging," the leaving of some rind tissue with the fruit stem attached to the twig when the fruit was pulled away, and such fruit spoiled rapidly in transit to market. Because of the losses thus caused, the industry turned almost entirely to clipping each fruit from its stem, a slower process but one which gave a much higher percentage of sound fruits. From 1900 to 1940, fruit clippers were used regularly, but shortage of labor during World War II brought about a return to pulling of oranges and grapefruit for faster harvesting. It was found that with suitable care in doing it so as to leave the "button" on the tree, pulling caused no more injury to the crop than usually occurred from clipper cuts or punctures.

Mandarin types "plug" easily when pulled, because of the nature of their rinds, and must be clipped, as may be necessary with 'Pineapple' oranges early in their season. By the time oranges and grapefruit are

FIG. 11. Bulk handling of fruit in picking operation in Lake County.
Courtesy Apshawa Groves, Inc.

well matured the fruit should separate readily when pulled properly. This means turning the fruit at right angles to its stem and exerting a sideways stress, not pulling straight away from the twig, for this latter technique may cause "plugging" even with very fully mature fruit.

Labor for harvesting the citrus crop has dwindled and its costs increased rapidly over the last decade. This has led to considerable research by experiment stations, commercial companies, and grower organizations directed toward mechanized harvesting methods. Studies have been, and are being, conducted along three lines: (1) development of best tree sizes, shapes, and spacings to facilitate effective equipment use; (2) discovery of abscission (growth regulator) compounds to reduce the pull force necessary to sever the fruit from a fruiting branch; and (3) invention of equipment to actually accomplish the harvest. Progress has been made. New plantings are spaced to anticipate hedging (pruning) practices. Cycloheximide, an abscission chemical, has been cleared for use on an experimental basis. Harvesting equipment (tree and limb shakers, blowers, water guns, and augers) reflect the innovative thinking of engineers and horticulturists. The concerted effort of the entire citrus industry is beginning to show results.

Knowing when to harvest citrus fruit involves the question of its degree of maturity. Properly speaking, there is no ripening process in citrus fruit and no such thing as "tree-ripe" fruit or ripe fruit of any kind. In peaches and avocados, as in most other familiar fruits, there is a stage of maturity at which softening takes place rapidly, and in a few days a hard fruit passes through this ripening process and becomes overripe. Citrus fruits pass from immature to mature and finally to overmature condition while remaining on the tree, but the changes are slow and spread over several months. When picked at any stage of maturity, the fruit does not change after picking except as it may be infected by fungi and spoil, or may slowly dry out. There is a considerable period of time during which the fruit has desirable quality, and during this period the fruit is considered properly mature.

Citrus growers want not only a large crop but one which will bring a good market price, which is based on internal and external quality. Internal quality is concerned with juiciness, flavor, and taste, while external quality reflects eye appeal. Although fruit for the processing industries does not need to have as high external quality as that for the fresh market, the internal quality must be as high or higher.

Determination of when any of the sweet citrus fruits have reached the maturity to be harvested to give a high-quality product for the consumer is an easy matter for the man who has only a few trees. He feels no pressure for getting his crop off the trees, and can sample fruits at intervals until he is satisfied that they have reached prime eating condition before he picks them in quantity for use by his family. If he wishes to ship a box of fruit to a friend, he will take pains to select only fruit of good quality and to pack it carefully so that it will reach its destination in prime condition. He needs no chemical tests to tell him when his fruit is mature, nor any regulations to assure that only fruit of acceptable quality is shipped.

It would be equally possible for every grower, large or small, to use the same criteria for determining maturity, but only in a utopian society could this assure good quality for the consumer. Too many growers proved long ago willing to ship immature fruit to distant markets, and as early as 1911 legislation was initiated to protect the reputation of Florida fruit. As the Florida citrus industry has increased in size and complexity more and more laws and regulations have been necessary to control the operations of grower, packer, and shipper for the best interests of the whole industry and to assure marketing of a product which the consumer could buy with confidence. These regulations deal with maturity, sizes, and grades of fruit which can be handled commercially, and are rather extensive, since it is necessary to set up different standards for each of the different kinds of citrus fruits. Each shipment of fresh fruit and each lot of fruit delivered to the cannery or concentrate plant is inspected to see that it meets the applicable legal standards.

What constitutes maturity in citrus fruits is defined by state law in each of the citrus-growing states, with a further federal statute which is applicable in case any state fails to enforce its law properly. Florida citrus fruits cannot be shipped commercially until they are mature as defined by Florida state laws. The Florida Citrus Commission of the Department of Citrus is authorized, under these laws, to issue regulations for determining maturity and for the limits of the various sizes and grades. The enforcement of these regulations is done by the Fruit and Vegetable Inspection Division of the Florida Department of Agriculture and Consumer Services. However, not all fruit which meets maturity requirements and is properly classified by size and grade can be shipped. To further insure orderly marketing, the citrus industry operates under five federal marketing agreements, voluntarily entered into. The first of these stipulates the grades, sizes, and containers of oranges, grapefruit, tangerines, 'Temples', and 'Murcott Honey' tangerines which may be shipped; three more apply to prorated shipments of interior grapefruit, Indian River grapefruit, and interior oranges; and the last regulates lime shipments. All are somewhat flexible, so that requirements may be made more stringent or be relaxed somewhat, in accordance with the nature of the crop being marketed.

Minimum Maturity Standards

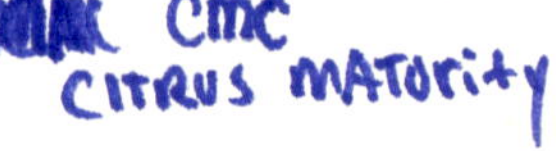

The first attempts to frame a legal definition of a mature citrus fruit were made in 1912. A single fruit character was sought which could be readily determined by physical or chemical methods and could be used to assure average consumer acceptance. This proved impossible in the case of most kinds of citrus. Only acid fruits (lemons and limes) can be judged suitable for shipment on the basis of a single character, per-

centage of juice by volume, and this is not related to degree of maturity. These fruits develop their maximum acidity early in the season while they are quite immature and are picked as soon as they meet minimum size requirements which are dictated primarily by market demand.

Limes are wanted only in a biologically immature state, hence legal maturity is defined only in terms of minimum size and juice content. The size requirement is to assure freedom from "green" taste and may change at times, but usually no 'Persian' limes less than 1¾ inches in diameter may be shipped at all. The general requirement is 42 per cent juice by volume, with a size restriction at or slightly below 1¾ inches as set by marketing agreement regulations each season. There is no size requirement for 'Key' limes but they also must have 42 per cent of juice by volume. No single character has been found to measure consumer acceptance of the non-acid citrus fruits; thus, a group of 5 interrelated characters must be employed. These are color break, juice content, soluble solids content, acid content, and ratio of solids to acid.

Color Break

Immature citrus fruit are green, like leaves, because their rinds contain large amounts of the green pigments, chlorophylls, which leaves contain. If temperatures are not too high, chlorophylls begin to diminish in quantity as the citrus fruits begin to mature, so that the yellow pigments, carotenoids, which are also present in green fruits (and leaves), become evident. Eventually the green color is likely to disappear entirely in fully matured fruits and is replaced by yellow in lemons and grapefruit, orange in oranges, or red in some mandarins and tangelos. This breakdown of chlorophylls takes place only when the temperature is relatively low, as in fall and winter in Florida. In the tropics, oranges are still more or less green when fully mature; and 'Valencia' oranges which have lost the green color in early spring may regreen if left on the trees into April or later in Florida.

Normally citrus fruits exhibit a change of color with maturity, and the first indication of change from leaf-green to a yellowish green, if it is developed naturally, is termed "color break." Florida law requires that citrus fruits (other than limes and lemons) show a color break to some extent before they can be shipped as fresh fruit. Treatment with ethylene gas in the "degreening" ("coloring") room in packinghouses greatly accelerates the breakdown of chlorophylls if the process has already begun naturally and makes it possible to develop satisfactory color for the market. Oranges, grapefruit, tangerines, 'Temples', and tangelos have a minimum requirement of color break all through the

season for fresh fruit, but there is no such requirement after December 1 for cannery fruit.

Temperature is not the only factor involved in color break, for light plays a part too. Fruit in the outside canopy will show this change earlier than that in the more shaded parts of the tree, thus early in the season it is often necessary to spot-pick exposed fruit. Sometimes, especially in a warm fall, shaded fruit may remain quite green after it has become satisfactorily mature internally. Satsumas, some tangerines, and tangerine hybrids, which are most highly prized when they have developed a deep red or orange color, may pass the condition of prime eating quality and be somewhat overmature before shaded fruit shows much color in the rind. 'Parson Brown', 'Hamlin', and some navel oranges may also be fully mature when still exhibiting some green rind color. Warm, rainy weather in the fall causes slow development of color break and duller color when the fruit is fully matured. The grower may also influence the development of color break by cultural practices, some of which (e.g., oil sprays after mid-July or excessive nitrogen fertilization) may delay this process, while others (e.g., properly balanced nitrogen with potash fertilization) may hasten it slightly.

Brightest rind color of citrus fruits is developed under conditions of bright sunshine, low atmospheric humidity, and relatively low temperature during the maturing period. Climatic conditions in Florida are not such as to cause high coloring of oranges, although internal fruit quality may be excellent. Therefore, after the degreening room treatment, they are dipped into a solution of orange-red dye to enhance their "eye appeal." Artificially colored fruits must be individually stamped "Color-Added" and must meet higher standards for internal quality than naturally colored fruits. 'Temples' and tangelos may also be so treated. This process is rigidly supervised to prevent fruit of unsatisfactory eating quality from being made to appear properly mature. Later in the season, as natural color develops in the rind, these fruits are usually shipped without coloring.

Juice Content

The juice sacs in citrus fruits become gorged with juice as the fruit matures. Good palatability involves a sufficient amount of readily released juice of desirable taste, and the laws specify minimum quantities which must be present. These are easily determined but are stated in different ways for different fruits. The required juice content of oranges is stated in terms of gallons of juice which can be squeezed from the fruit in a standard packed box, with the minimum requirement of 4½ gallons for "natural-color" oranges and 5 gallons for "color-added" fruit. Juice content of grapefruit is stated as the cubic centimeters per fruit, with a minimum quantity for each standard size. No juice content is specified for mandarins, 'Temples', tangelos, or other hybrids. As mentioned previously, the juice requirement for limes and lemons is stated as a percentage of their volume.

The high juice content characteristic of Florida citrus fruits, especially of oranges, is an outstanding factor in their reputation. Early or late in the season some fruits may have less than the minimum, but as a general rule Florida citrus fruits run far above the minimum requirement for juice. Sometimes the pink and red seedless grapefruit varieties have difficulty in meeting the requirement early in the season. A generally helpful legal requirement may sometimes do minor injustice.

Soluble Solids

The earliest minimum maturity standards set up in Florida were based on percentage content of sugar, determination of which required considerable time as well as chemical equipment. Later it was found that sugars constituted quite regularly about 75 per cent of the total of the compounds—organic and inorganic—dissolved in the juices, the so-called soluble solids. These soluble solids can be measured easily by determination of the density of the solution, and thus the change in sugar content is easily measured too. There is a steady increase in soluble solids, and of sugars proportionately, all during the season until the fruit is well past acceptable maturity. In addition to sugar, the soluble solids include certain organic acids (largely citric acid), vitamins, and very small amounts of inorganic salts and organic compounds (fruit esters) which are responsible for the flavors characteristic of the various citrus fruits. The off-flavors which sometimes develop in fruits because of improper handling or overmaturity are usually due to changes in these flavor compounds, unless microorganisms have invaded the fruit.

Soluble solids begin to accumulate appreciably in early fall and reach levels of desirable palatability in the fall, winter, or spring, depending on the variety involved. Density of the juice is measured with a hydrometer floating in a cylinder of juice, and the reading of this hydrometer in degrees Brix is converted by tables, after correction for temperature, into percentage of soluble solids. Although late in the season of any variety the values may be higher, the following figures give the usual range of total soluble solids during the normal harvesting season for the important citrus fruits: oranges, tangerines, 'Temples', and tangelos, 9 to 14 per cent; 'Murcott Honey' tangerine, 12 to 16 per cent; grapefruit, 7 to 12 per cent. Minimal values of solids are required for shipping these fruits.

Acid Content

Citric acid is the one chiefly found in citrus fruits, but small amounts of malic, oxalic, and tartaric acids are also present. Acid percentage is highest in quite immature fruit and decreases steadily during the maturing period. There is a gradual change in taste from sour to pleasantly tart to sweet with no consciousness of acidity, and then very late in the season an insipid sweetness for lack of enough acidity to make the taste pleasant.

There is no maximum permissible acidity specified for any citrus fruit, but the percentage must always be determined in order to calculate the required ratios. Minimum acid content is required for oranges (0.4 per cent if naturally colored, 0.5 per cent if artificially colored) and tangelos (0.4 per cent), but not for other citrus fruits. The acidity is determined by titration of the juice with a standard solution of a base; the equivalent acidity as anhydrous citric acid can be calculated easily from the quantity of base used. The amounts usually found in sweet citrus fruits during their normal harvesting season range from 1 per cent early in the season to 0.5 per cent late in the season. Limes and lemons, which are valued for their high acidity, have from 4 to 7 per cent of acid, according to variety.

Ratio of Total Soluble Solids to Total Acid

While the flavors of the different citrus fruits are chiefly due to small amounts of certain organic compounds, as explained previously, the taste (sweet or sour) is dependent on the relative amounts of sugar (determined as soluble solids) and acid in the juice. Extensive research into consumer preferences for several sweet citrus fruits, using the technique of "taste panels," has shown that no particular sugar content will alone assure acceptable taste. The average consumer wishes a pleasantly tart or subacid fruit, and this requires certain proportions of sugar to acid. The percentage of total soluble solids divided by the percentage of total acid gives the ratio of solids to acid, always stated on the basis of acid as 1. Data from thousands of taste tests with fruits of all degrees of maturity made possible the charting of what constituted satisfactory ratios for a wide range of solids and acids, and from these findings the present laws were drawn.

Early in the season of any fruit the solids are low and acid high, giving a low ratio. As the season advances the solids increase and the acid decreases, giving progressively higher ratios. The laws require that fruit early in its season must have a certain minimum content of soluble solids and minimum ratios of solids to acid. For grapefruit these are 7.0, 7.5, and 8.0 per cent solids, for seeded, white seedless, and pink and red seedless, respectively, and a 7.0 to 1 ratio; for oranges, 9.0 per cent solids and 10.0 to 1 ratio if natural color and 9.2 per cent solids and 9.9 to 1 ratio if "color-added"; for tangerines and 'Temples', 9 per cent solids and 9.0 to 1 ratio; and for tangelos, 9.5 per cent solids and 9.75 to 1 ratio. A higher ratio is necessary to give acceptable taste when the sugar content is low than when it is high, so if fruit exceeds the minimum solids requirement, it may have a lower ratio. Sugars increase as the fruit becomes more mature, hence the ratio required decreases as the season advances. Thus, oranges with 11 per cent solids need only have a ratio of 9.0 to 1. Acid content decreases with increasing maturity so that the same ratio is attained with lower solids and the minimum solids requirement decreases as the season advances. Although natural color oranges must have 9.0 per cent solids until the end of October to be shipped, they need have only 8.5 per cent solids after November 15.

Citrus fruits just approaching satisfactory maturity often have an unpleasant, raw taste which disappears as they become fully matured and which cannot be measured by simple chemical tests. Higher ratios early in the season act to decrease the possibility that "green" tasting fruit will reach the market. On the other hand, fruit with very high sugar content at the end of the season may be insipid. The minimum ratio specified (9.0 to 1 for oranges with 11 per cent solids) assures that no such fruit will be marketed.

In passing, it may be noted that a certain ratio is not in itself a guarantee of quality unless tied to total soluble solids. Oranges with 6 per cent solids, which is very low for oranges, but only ½ per cent acid would have a 12 to 1 ratio but very poor quality.

In summary, legal maturity for Florida citrus fruits is defined by laws and regulations which require a color break, minimum juice content, minimum percentages of total soluble solids and acid and required solids-acid ratios. A lot of fruit must meet all of the applicable requirements, as shown in tests by federal-state inspectors, before it is certified for handling or shipping. Minimum requirements for legal maturity of Florida citrus fruits have become higher and more complex through the years, and it can be said with pride that the interests of the consumer have been a concern of the men who lead the industry which produces and markets these delectable fruits. Emphasis is now placed on market restrictions as "minimum quality standards," rather than purely legal requirements.

Sizes and Containers

The market sizes of citrus fruits correspond to the number of fruits of uniform size which could be packed by set patterns in a standard shipping box of 4/5 bushel capacity. Each size is defined by the maximum and minimum diameters of the fruit in it, since there is a small range of diameters for each size. On entering a packinghouse for sweet citrus fruits, which are more or less spherical, it is easy to recognize the "sizers" which are the machines on which the fruits move along slightly tapered rollers until they reach the proper diameter range and drop between the rollers into the appropriate size bin. Limes are longer than they are wide and are not so easily sized in this way; more often the fruits are sized by weight as they move over a series of counterbalanced trap doors, the heaviest fruit dropping out first.

In citrus fruit there is no relation between size and grade, as there is in apples. Certain sizes are more popular with the housewife than others, however, particularly those in the middle of the range of sizes for any kind of citrus fruit. Extra large sizes may be less juicy than average, are more difficult to squeeze juice from, and cost more per fruit; fruits in the small sizes are usually higher in juice and sugar than the average, but more of them must be squeezed to get the same volume of juice. The preference for fruits of medium size is well known to all fruit handlers and is recognized by the Marketing Agreement committees, which may restrict the sizes as well as the grades of any citrus fruit which may be marketed. If the crop is running heavily to small sizes, then smaller sizes will be permitted to go to market than when the bulk of the crop is of medium or large sizes.

Legal market sizes vary for the different kinds of citrus fruits and are defined in United States standards as regulated and amended by the Florida Citrus Commission. The standard nailed 1 3/5 bushel box for which sizes were originally established was formerly used almost exclusively for shipping citrus fruits, but is no longer in use. Commercial shipping sizes are 48, 64, 80, 100, 125, and 163 for oranges, 'Temples', navel oranges, and tangelos, and 18, 23, 32, 40, 48, and 56 for grapefruit in regular 4/5 bushel wirebound boxes or cartons. Sizes for tangerines in the 4/5 bushel flat, wirebound box are 100, 120, 150, 176, 210, 246, and 294. Cartons of 2/5 bushel size, and 5 and 8 pound polyethylene and plastic mesh bags are also used. Fruit in bags is sold by weight; hence, bags may include more than one size of fruit. Limes are not packed by standard sizes, although fruits in any one container must be uniform in size within a range of ¼ inch in diameter.

Grades

U.S. standards for grades of Florida citrus fruits are defined by the Consumer and Marketing Service of the United States Department of Agriculture. Florida standards for grades, which apply to citrus fruits shipped intrastate, are similar in most respects to the federal standards. They are established under regulations of the Florida Citrus Commission. Regulations of the several federal marketing agreements issued by the U.S. Secretary of Agriculture upon the recommendation of the Growers Administrative Committee specify what grades and sizes of citrus fruits may be shipped outside of the production area (south of Georgia and east of the Suwannee River). These regulations are established at the beginning of each shipping season and are revised or amended as crop and market conditions warrant during the season.

The grade for a citrus fruit is based mostly on external appearance and represents the degree of satisfactory appeal to the eye of the consumer. However, no fruit may be placed in a grade unless it meets minimum requirements for maturity.

Four grades are commonly recognized: U.S. Fancy, U.S. No. 1, U.S. No. 2, and U.S. No. 3. Fruit which has not yet been graded is tabulated on packinghouse records as Unclassified. The U.S. Fancy grade is seldom packed since it represents fruit of external perfection very difficult to produce. Most citrus fruit moves as U.S. No. 1, and the grower attempts to practice grove care which will enable 60 per cent or more of his crop to be packed in this grade. Occasionally he can also ship some which grades U.S. No. 2. Fruit of U.S. No. 3 grade is rarely shipped to fresh-fruit markets, but is sent instead to the processing plants.

To be classed in any grade, citrus fruits must be mature, similar in varietal characteristics, and free from unhealed cuts and bruises. The Fancy grade requires that fruit be well colored, well shaped, firm, fairly smooth in texture of rind, and free from blemishes on the rind due to pests or mechanical causes, except that up to an average of 10 per cent of the surface of any lot of fruit may be discolored by rust mites. These represent the external characters which the housewife considers to indicate highest quality, and the various lower grades specify the extent to which fruit may progressively vary from these ideals.

The U.S. No. 1 and U.S. No. 2 grades are further divided into subclasses based on discoloration due to rust mites. Some people suppose that russetted fruit is a different variety from bright fruit, and others believe it to be sweeter than bright fruit. As mentioned in Chapter 9, rust mites affect only the appearance of the rind and do not have any effect on internal quality of fruits on which they feed. Russetted fruit is just as good to eat as bright fruit, and some markets even pay a premium for it, although it is usually discounted somewhat for poor external appearance and may not keep as long.

The subclasses are separated on the basis of the proportion of the rind which is discolored and the percentage of fruit showing russetting. U.S. No. 1 grade has five subclasses: U.S. No. 1 Bright, U.S. No. 1 (no subclass name), U.S. No. 1 Golden, U.S. No. 1 Bronze, and U.S. No. 1 Russet; while U.S. No. 2 has extra Bright and Russet subclasses. All these subclasses have the same grade characters except for rind color, and represent progressively greater amounts and intensity of discoloration. U.S. No. 1 Bright may contain no fruit with more than one-fifth of the whole area of its surface discolored, while U.S. No. 1 Russet must have at least a third of the fruits (by count) with more than one-third of the surface discolored. The discoloration becomes darker as the affected portion of the rind becomes larger. Sometimes the discolored area is dull and will not take a shine (early season damage), which is undesirable; in other cases this area is glossy, although discolored, and will take a shine from the polishing brushes, which is desirable from the marketing standpoint.

U.S. grades for 'Persian' limes are U.S. No. 1, U.S. No. 2, and U.S. Combination, which contains at least 60 per cent of U.S. No. 1 grade. Each grade requires fruit to be mostly green in color, but has subclasses, Turning and Mixed Color, for fruit which is becoming yellow. The minimum grade which may usually be shipped is U.S. Combination Mixed Color, although this requirement is subject to change upon the recommendation of the Lime Administrative Committee in accordance with the needs of the industry as indicated by the current crop condition. 'Key' limes must meet standards for U.S. No. 2 'Persian' limes except for no color specification.

In summary, maturity denotes a certain minimum of acceptable edibility, size refers to the number of fruits which can be packed in standard boxes, and grade indicates a certain degree of acceptability of external appearance.

Fruit for Processing

The discussion thus far has been centered on fruit which is to be shipped to market as fresh fruit, since sizes and grades are primarily applied only to such fruit. The requirements of the citrus processor must also be considered, however, since about 85 per cent of the total crop now goes to processing plants. This is not equally true of all types of citrus fruits. In the 1971–72 season over 90 per cent of sweet oranges were processed (mostly to frozen concentrate); two-thirds of

TABLE 6

UTILIZATION OF FLORIDA CITRUS FRUITS
(*in 1,000 boxes*)

Type of Citrus Fruit	1969–70 Fresh Sale	1969–70 Processed	1970–71 Fresh Sale	1970–71 Processed	1971–72 Fresh Sale	1971–72 Processed
Orange, Early and Midseason	7,491	65,409	8,198	73,902	6,024	62,776
Orange, Late	5,772	59,028	5,764	54,436	5,209	62,991
Grapefruit, Seedy	559	8,941	479	11,321	285	10,615
Grapefruit, White, Seedless	7,016	10,684	7,452	12,748	8,938	14,862
Grapefruit, Pink, Seedless	6,687	3,513	7,029	3,871	7,816	4,484
Tangerines	2,383	617	2,661	1,039	2,228	972
Temples	2,823	2,377	2,232	2,768	1,783	3,517
Tangelos	1,397	1,103	1,607	1,093	1,896	2,004
Limes	374	351	395	485	474	626

Source: Florida Agricultural Statistics, Citrus Summary, 1972

the grapefruit went to processing plants; and 32 per cent of the tangerine crop was processed. 'Temples', 'Murcotts', and tangelos were processed in relatively small amounts.

Internal quality is of greater importance to the processor than external appearance, but when there is a heavy crop and the fresh-fruit market is slow, the processor tends to discount fruit for external quality even though the lower-grade fruit is fully as satisfactory for his use. When fruit is in short supply the processor is glad to overlook external appearance unless there are rind blemishes which will flake off into the juice or the fruit is so abnormal in size or shape that it cannot be handled by the processing machinery. As fruit begins to dry out, either through natural processes as it becomes overmature or because of freeze injury, the value to the processor decreases sharply. Any fruit with less than average juice content is penalized.

Fruit is sold to canneries on the basis of pounds-solids (pounds of juice times percentage of total soluble solids) or pounds-juice per box or load, the latter for single-strength (canned) juices. Random samples are taken for fruit quality (maturity) tests from each load upon arrival at the cannery. Information furnished the processor not only includes the usual figures on Brix (total soluble solids), acids, and Brix-acid ratio obtained from a regular maturity test to assure compliance with legal requirements, but also the number of box equivalents or weight of fruit in the load and the weight of the juice per box equivalent. These data enable the processor to make necessary calculations as to probable yield of final product and how the load should be blended with others for efficient plant operation.

Citrus canneries in Florida process a multitude of products in six principal forms—frozen, chilled, and canned juices and frozen, chilled, and canned sections and salads—plus numerous valuable by-products, including peel oils, molasses, dried pulp, etc. About 80 per cent of all of the fruit delivered to processors goes into two products, frozen concentrated orange juice and chilled juice including orange-grapefruit blends. Fresh orange juice must have a minimum Brix-acid ratio of 10 to 1 and reconstituted juice from 13 to 1 to 19.5 to 1 under U.S. and Florida standards for frozen concentrated orange juice. The product is manufactured principally as either a 3 plus 1 or 4 plus 1 concentrate. A can of the former has a Brix of from 45° to 47° as purchased in the grocery store. Fresh juice is concentrated in a multistage high vacuum process to about 55° Brix, after which fresh juice is added back to reduce the Brix to the proper range. Addition of fresh juice (and oil flashed off during concentration) is essential to produce a high quality product. Other frozen concentrated juices include grapefruit, tangerine, lemon, and lime. An increasing proportion of the oranges and grapefruit sent to canneries is being processed as chilled juice, with over 20 million boxes of oranges and nearly 3 million of grapefruit being so utilized in 1972–73. Juice from fruit of concentrate quality is extracted, flash pasteurized, chilled to near freezing and put into containers ranging from tank cars or ships to 4 oz. cartons for shipment to market where it is sold in grocery stores or delivered to individual homes on a milk route. Widespread acceptance of frozen concentrated and chilled juices has led to an increasing number of various types of sections and salads being made available in either frozen or chilled form. A diminishing proportion of fruit is being processed into canned products, with consequent lower returns to the grower whose fruit will not meet concentrate standards and must be diverted to single-strength juice.

Today, it is recognized by knowledgeable growers that virtually

TABLE 7

PROCESSOR UTILIZATION OF FLORIDA CITRUS FRUITS
(in 1,000 boxes)

Process	1970–71	1971–72	1972–73
Canned			
Grapefruit			
Juice	15,245	15,068	13,884
Sections and Salads	2,434	2,053	2,149
Orange			
Juice	8,531	7,118	8,693
Sections and Salads	91	98	72
Tangerine			
Juice	25	11	18
Concentrated			
Grapefruit	6,819	8,717	8,212
Orange	103,551	104,362	132,060
Tangerine	1,000	1,089	943
Chilled			
Grapefruit			
Juice	2,347	3,202	2,901
Sections and Salads	1,091	993	1,206
Orange			
Juice	19,721	19,207	20,382
Sections and Salads	704	532	652

Source: Florida Canners Association

every phase of the cultural program from the initial choices of a rootstock and site to fertilizer, spray, and other components of the program has an influence upon fruit qualities and yield. The grower of fruit for processing has, moreover, a convenient yardstick in pounds-solids per acre against which he may measure the overall efficiency of his operation. Most growers realize, however, much as they would wish to sell all of their produce to a single outlet, that a healthy, viable fresh fruit market is also essential to their success in supplying fruit profitably to processors. Their well-being is assured as long as the housewife continues to identify the juice she reconstitutes from a small can, or pours from a carton, with freshly-squeezed juice or cut-up fresh fruit. There are many synthetic citrus or citrus-like products around ready to become substitutes should this image ever be lost.

12. Citrus Fruits for the Home

ONE OF THE advertisements during the Florida land boom of the 1920's exhorted the prospective purchaser to buy a home and "pick oranges from your piazza." While this is a little too much to expect, certainly one of the satisfactions which nearly every homeowner in Florida can have is to be able to pick some kind of citrus fruit from his own trees.* The number of different kinds which may be so grown will vary with size of the home grounds and with the climatic conditions of their location.

As one goes from north to south in the state, the number of varieties which are climatically adapted increases steadily. In the panhandle of northwestern Florida, only the kinds hardiest to cold are suitable—satsumas, kumquats, calamondin, and the like. In northern peninsular Florida, sweet oranges, grapefruit, tangerines, and 'Meyer' lemon may be added, especially in places with cold-protection afforded by lakes or buildings. In the Leesburg-Orlando area, almost any kind of citrus may be grown except limes and citron, and south of Tampa-Haines City all varieties are suitable. Occasional freezes may injure trees, especially the more tender kinds, as here located, but as a general rule it is reasonable to expect satisfaction on this basis.

Considering the various types of citrus fruits in turn, the citrus industry lists them as staples (sweet oranges and grapefruit) which are accepted as a regular part of the diet, eaten in some form practically

*Three types of use are envisaged for citrus trees in the home grounds: to supply the family with fruit, to provide fruit for shipping as gift boxes, and to beautify the landscape.

every day by many people; specialty fruits (tangerines and tangerine hybrids) which are excellent for holiday or dessert uses; and acid fruits (lemons, limes, and others with high citric acid content) which find use as thirst-quenching drinks, garnishes on the dinner table, and ingredients for refreshing pies and delicious cakes. For landscaping the homesite, the dual-purpose fruit trees of ornamental value enhance the beauty of the surroundings, while a specimen tree of some exotic type adds interest as a "conversation piece." Depending upon the size of the homesite, several varieties of each type may be selected.

Sweet orange is the first choice, and if only a single citrus tree is to be grown, it may well be an early variety such as 'Hamlin', except where the danger of cold makes it necessary to substitute 'Owari' satsuma. If there is room for more than one sweet orange tree, a midseason variety such as 'Pineapple' would be the second one, and the late 'Valencia' would be the third. In the area north of Ocala, however, 'Valencia' is not recommended because the fruit is too often injured by cold. Where 'Valencia' can be grown successfully, these three varieties will supply fresh fruit continuously from early November to August.

Grapefruit is usually second choice in citrus fruits. The 'Duncan' variety is one of the best for home use, its quality compensating for its seediness. Many people will prefer the flavor of 'Triumph', however, and some may choose a red-fleshed variety like 'Ruby' because of the combination of flesh color and seedlessness. Usually a single grapefruit tree will suffice whereas several sweet orange trees are wanted.

Among the mandarins, satsumas (usually a substitute for sweet oranges) mature during late November but hold quality on the tree for only about six weeks. The new mandarin hybrids ('Lee', 'Osceola', 'Robinson', 'Page' and 'Nova') should have much interest as early maturing tangerine substitutes. 'Robinson' and 'Nova' are now favored. 'Dancy' and 'Ponkan' tangerines may reach fair eating quality by Thanksgiving, but 'Dancy' is usually not well matured by then. For the holiday season the 'Ponkan' is superior in quality to 'Dancy', but does not carry as well or look so festive in gift boxes. 'Oneco' provides a tasty tangerine over several months, and for late spring the 'King' mandarin is of unsurpassed quality though not handsome in appearance. The homeowner may well decide to substitute a tangor or tangelo for a tangerine when choice must be limited. 'Temple' or 'Murcott' tangors are superb in quality as well as appearance, and so is 'Minneola' tangelo; and many people would choose one of these in preference to grapefruit if a choice were necessary. 'Minneola' needs a tangor, tangerine, or sweet orange to assure pollination for a good crop.

Acid fruits include lemons and limes, but even where cold permits them to be grown the true lemon is not recommended for a home fruit because of its susceptibility to scab. For southern Florida the 'Persian' lime makes a satisfactory acid fruit for the home, since fruit can be picked any time of the year. The 'Meyer' lemon and 'Eustis' limequat can be grown wherever sweet oranges can be, and the calamondin is a good acid fruit for home use anywhere in the state.

Dual-purpose fruits are those which combine ornamental value with useful fruit. The kumquats, calamondin, and 'Meyer' lemon are notable examples. Conversational fruits might cover such items as 'Ponderosa' lemon, 'Ruby' blood orange, shaddock, 'Thornton' tangelo, and perhaps a citrangequat or citrangedin. These fruits always excite interest on the part of guests.

Stocks should be chosen for hardiness, long life, and fruit quality rather than primarily for yield. Wherever satsumas are grown for hardiness to cold, they should be on trifoliate-orange stock. In northern Florida other citrus fruits may be grown on either trifoliate or sour orange as stocks. In central Florida sweet orange on well-drained soils and 'Cleopatra' mandarin on poorly drained soils will probably give the best fruit quality without risk of tristeza. Any citrus trees bought for the home planting should be certified free of the common virus troubles—psorosis, tristeza, exocortis, and xyloporosis—as well as free from nematode infestation. As pointed out in Chapter 7, the homeowner who wants citrus trees should also be very careful not to plant ornamentals which may be infested with nematodes and may render futile his caution in choosing nematode-free citrus trees.

Any well-drained site which is suitable for building a home and having a garden is likely to be satisfactory for citrus trees if temperatures are not limiting. These trees will tolerate light shade but will be more productive if they are not shaded by other trees. They should also not be planted so close together that they cut off light from the lower limbs of each other. The spacings recommended for commercial plantings should be exceeded for home plantings, not decreased.

Vigorous 1-year-old nursery trees should be planted, and preferably should be container-grown or balled-and-burlapped. These will cost more than the bare-rooted trees used in grove planting, but they will be much more certain to live and thrive with minimum attention. For the man who is planting only a few trees this assurance and relief from care is usually worth the difference in cost. These trees may be set out at any time when they do not have tender new growth, but it is unwise to plant before January 15 in southern Florida or February 1 in the northern part of the state. They should be set slightly higher than they grew in the nursery row or container, and be provided with a basin to hold water around each tree. This basin should be about 2 feet across and be filled with leaves to conserve moisture and keep summer soil

FIG. 12. Citrus trees around the home make for more bountiful living.
Courtesy Lake Region Packing Association

temperatures low. If subfreezing temperatures are forecast after trees are planted, they should be protected by covering them with bean hampers or by banking as described in Chapter 5.

Water is the first requirement of the young tree, and will be used up rapidly by trees with a good head of foliage, as container-grown trees will have. If a garden hose reaches the trees easily, let it run in each basin for 10 or 15 minutes each week, long enough to fill the basin pretty well with water, for the first month. Thereafter, whenever new growth is seen to wilt in midafternoon, fill the basins again. If a hose cannot be used, supply about 8 or 10 gallons at each watering. Need for irrigation will be less as the trees increase their root systems each year, but some occasions for watering may be expected through the fourth year.

Fertilization should start when swelling buds indicate that growth is beginning. The first year it is well to apply fertilizer about every six weeks from early April until September. A 6-6-6 analysis is a good general one for garden use, including citrus trees, and may be used at rates from ½ cupful for the first application to 1½ pints in September, increasing the quantity steadily all season. For the second, third, and fourth years, the fertilizing schedule given in Chapter 6 can be followed. Fertilizer should cover the basin area the first year. In succeeding years a good rule is to spread the fertilizer as many feet beyond the drip of the canopy as the age of the tree in years. For the home garden it is convenient to remember that 1 pint of mixed fertilizer weighs 1 pound.

Citrus trees in home gardens are much less likely to suffer from deficiency of mineral elements than in commercial groves, because of the greater amount of organic matter usually present in garden soils. As a precaution against the possible development of deficiencies, however, it is wise to use during the first few years fertilizer mixtures containing minor elements. These are readily available at fertilizer stores. If zinc deficiency symptoms appear, it will be necessary to apply a spray as described in Chapter 9, but this need is rather rare in gardens on sandy soils. On alkaline soils, however, zinc and manganese deficiencies may be expected to appear unless annual nutritional sprays are applied.

Many home gardeners have grown citrus trees successfully without any attention to soil reaction. If the garden is on sandy soil and sulfur is not used in a regular program of pest control, the gardener is justified in not worrying about this matter. If the trees do not have healthy foliage, however, and the cause is not obviously an insect infestation, it would be well to take a sample of soil to the County Extension

Director for checking the acidity. If the soil is not in the pH range of 5.5 to 6.5, the situation should be corrected as recommended or as explained in Chapter 8. If the soil is naturally basic, it will be impossible to change the reaction, but use of a heavy organic mulch has additional advantage in helping counteract the alkalinity of the soil by providing a more favorable medium for some of the roots near the soil surface.

Fertilization of bearing citrus trees in the home grounds is best based on age rather than yield because of the difficulty of keeping yield records. The table of fertilizer applications on p. 168 will be quite satisfactory for the home gardener to follow. However, these amounts are for trees growing without competition of lawn grass. Citrus trees in the home garden will grow best if they are cleanly cultivated or else kept heavily mulched with grass, leaves, or pine straw, but sometimes they are planted in the midst of an expanse of green lawn. In such cases the turf should be kept from growing closer than 3 feet from the trunk of the tree, and in the first ten years the amount of fertilizer should be increased about 25 per cent at each application to provide enough for grass as well as tree. If grass clippings are not carried away but are left to decay in place, there will be a return to the soil of the elements needed by the grass thereafter, and the trees will no longer have competition. Fertilizer applied for tree use will injure the grass if it is not promptly washed down to the soil with water. Under a mulch system, weeds should be suppressed by the mulch and there is no occasion to cultivate then.

Pest control should be undertaken only as need for it becomes evident. Citrus trees in the home garden may thrive for years with little trouble from pests. When they do appear in sufficient intensity to require action, the homeowner will usually find it more satisfactory to have the spraying done by a pest control operator or a neighboring commercial grower rather than to spray the trees himself. This latter may be perfectly feasible with trees up to bearing size, but larger trees require spraying equipment of larger capacity and power than is likely to be profitable for the grower of a few trees to own. The discussion of pests and their control in Chapter 9 is just as valid for the home garden citrus grower as for the commercial operator. However, the dangerous qualities of some insecticides are more important to the home gardener, since he is less likely to have adequate protective equipment for application and his house and family are more likely to be affected, than to the commercial operator. Parathion in particular should not be used on citrus trees in the home grounds.

Appendix

Tables of Weights and Measures
U.S. Standards to Metric System

Linear Measure

1	inch			=	2.54	centimeters
12	inches	=	1 foot	=	0.3048	meter
3	feet	=	1 yard	=	0.9144	meter
16½	feet (5½ yards)	=	1 rod	=	5.029	meters
5,280	feet (1,760 yards)	=	1 mile	=	1,609.3	meters

0.3937	inch	=	1 centimeter
39.37	inches	=	1 meter
0.621	mile	=	1 kilometer

Land Measure

1.0	acre	=	0.4047	hectare
2.471	acres	=	1.0	hectare

Liquid Measure

1 gill	=	4 fluid ounces	=	7.219	cubic inches	=	0.1183 liter
4 gills	=	1 pint	=	28.875	cubic inches	=	0.4732 liter
2 pints	=	1 quart	=	57.75	cubic inches	=	0.9463 liter
4 quarts	=	1 gallon	=	231.00	cubic inches	=	3.7853 liters

Dry Measure

1 pint			=	33.60	cubic inches	=	0.5505 liter
2 pints	=	1 quart	=	67.20	cubic inches	=	1.1012 liters
8 quarts	=	1 peck	=	537.61	cubic inches	=	8.8096 liters
4 pecks	=	1 bushel	=	2,150.42	cubic inches	=	35.2383 liters

Weight

1	ounce			=	28.3495	grams
1	pound			=	453.59	grams
100	pounds (hundredweight)			=	45.36	kilograms
2,000	pounds (1 ton)			=	907.18	kilograms
2.2046	pounds	=	1,000 grams	=	1.0	kilogram

Temperature

The Fahrenheit thermometer is a thermometer on which the boiling point of pure water is 212° and the freezing point 32°, under standard atmospheric pressure, abbreviated F.

The Celsius thermometer (formerly Centigrade) is a thermometer on which, under laboratory conditions, 0° is the freezing point and 100° the boiling point of pure water, abbreviated C.

Some equivalents:	°F	°C	
	212	100	Boiling
	194	90	
	176	80	
	158	70	
	140	60	
	130	54.5	High endurance range for citrus (130–100)
	122	50	
	104	40	
Growth range for citrus (100–55)	100	37.75	
	86	30	
	68	20	
	55	12.75	Low endurance range for citrus
	50	10	
Low lethal range for citrus (32–20):	32	0	Freezing
Threshold temperatures:*	28	– 2.22	
Tender growth: 32°F.	24	– 4.44	
Fruit and terminal growth: 28°F.	20	– 6.66	
Scaffold limbs: 24°F.	10	–12.22	
Trunk: 20°F.	0	–17.75	

* Threshold temperatures are approximated for commercial practice. Damages will vary with species and variety of stock and scion, health and stage of growth of the plant system, duration of the low temperature, and atmospheric conditions preceding, during, and following period of low temperature.

Index

abscission chemicals, 191–92
acid content of fruit, 224
acid soils, 96, 98, 99, 163–64
adaptability of stocks, 65
air drainage, 85, 90, 101, 112
albedo, 13
aldrin, 181
'Algerian' navel orange, 150
alkaline soils, 98–99, 100, 163–64
ant control, 134
anthracnose, 187
aphids, 146, 179
Aphytis wasps, 174, 188
application of nutrient elements: on ground, 166–69; in sprays, 169–71
areas of production in Florida, 103–10; new grouping: East Coast, 108; Lower Interior, 108; Upper Interior, 108; West Coast, 108; old grouping: Indian River, 86, 106–7; Lower East Coast, 106; Lower West Coast, 106; North Central, 106; South Central, 106, 107; Upper West Coast, 106
arsenical sprays, 189–91
Astatula fine sand, 96
Aurantioideae, 12–14
available water of soil, 102, 200–201
'Avon' lemon, 48

banking: budded seedlings, 77; newly planted trees, 125; removing banks, 127; young trees, 134–35
bearing grove: basic factors in production, 136–38; rehabilitation, 138–42
'Bearss' lemon, 47; lime, 51
beneficial fungi, 188–89; insects, 187–88; mites, 188; snails, 189
biological control of pests, 187–89
black scale, 178
blood oranges, 22, 23, 31, 32
boron as a nutrient, 162, 171
broad mite, 184–85
brown rot, 187
budding: after-care, 77; methods of, 75–76; selecting budwood for, 71–72; selecting parent tree for, 71; top-working by, 79
budlings, care of, 77–78
'Burgundy' grapefruit, 37–38
burrowing nematode, 148–51

calamondin, 52–53, 61
calcareous soils, 98, 99, 100, 163–64
calcium: as a nutrient, 159; as a soil amendment, 165–66
canker: bacterial, 185; frost or cold, 88
care of young trees: ant control, 134; cold protection, 134–35; cultivation, 133–34; fertilization, 128–30; pruning, 132; spraying, 130–31; unbanking, 127; watering, 127–28
'Carrizo' citrange, 61, 70, 142, 150
chaff scale, 177
chelates, 160, 163
chemical weed control, 198–99
'Chinese Honey' orange, 43–44
chlorobenzilate, 182–83
chlorosis, 154–55
citrange, 16, 61, 70, 147
citrangedin, 61–62
citrangequat, 61–62
citron: culture of, 53; exocortis, 147; first known, 2; origin of, 2

citrus: diseases, 142–51, 185–87; hybrids, 14, 53–62; insect pests, 173–81; origin of word, 3; processing, 20–21; uses, 20–21
Citrus: as commercial genus, 13; *aurantifolia,* 50; *aurantium,* 67; characters common to all species, 19–20; *jambhiri,* 49, 66–67; *latifolia,* 50; *limetta,* 52; *limettoides,* 52; *limon,* 45; *limonia,* 45, 53; *macrophylla,* 142–43, 147, 148; *madurensis,* 52; *medica,* 53; *meyeri,* 48; *microcarpa,* 52; *mitis,* 52; *nobilis,* 39; origin of species, 1; *paradisi,* 31, 70; *reshni,* 44, 67–68; *reticulata,* 38–44; *sinensis,* 22, 68–69; *tangerina,* 43; *trifoliata,* 14; *unshiu,* 39
citrus bud mite, 185
Citrus Budwood Registration Program, 71
citrus canker, 185
citrus fertilizer program, 164
citrus mealybug, 178
citrus nematode, 151
citrus production: in Florida, 7, 8; in other countries, 9–11; in other states, 8, 9
citrus red mite (purple mite), 183
citrus root weevil, 181
citrus rust mite, 182–83
citrus snow scale, 176
'Cleopatra' tangerine, 44; as stock, 67–68, 142, 144
climate in citrus growing, 83–94
Clymenia, 13
cold hardiness of citrus species, 20, 88–89; as affected by stocks, 89
cold injury: dormancy and, 87–89; duration of cold and, 88; effect on costs and returns, 216; evidences of, 112–13, 214–16; previous conditions and, 90; pruning after, 206, 215–16; shade and, 90; temperature and, 84–86; to fruit, 214–15; windbreaks and, 86, 90
cold protection. *See* protection from cold
color-added, 222, 225
color break, 221–22
copper: as a fungicide, 185–87; as a nutrient, 160–61, 164
cottony-cushion scale, 177
cover crops, 194–96
crop, the: fruit sizes, 226–27; grade standards, 227–28; harvesting, 217–19; maturity requirements, 220–26; processing requirements, 229–31; utilization, 229–30
cultivation: of bearing trees, 193–98; of young trees, 133–34

damping-off of young seedlings, 187
'Dancy' tangerine, 43
DBCP, 151
deadwood, pruning of, 205–6, 215
'Delicious' orange, 23, 30
Delnav, 183
dieldrin, 181
Diplodia root rot, 145
diseases causing tree decline: foot and root rots, 143–45; nematodes, 148–51; viruses, 145–48; young tree decline, 66–67, 142–43
diseases of fruit: bacterial, 185; fungal: brown rot, 187; greasy spot, 186–87; melanose, 185–86; scab, 186
dolomite (dolomitic limestone), 165
dormancy and cold injury, 87–89
"dormant" budding, 75
double planting, 118, 119
'Dream' naval orange, 28
dual-purpose fruits, 234
'Duncan' grapefruit, 36

embryos, nucellar, 14
'Enterprise' orange, 30
Eremocitrus, 12
essential elements. *See* nutrient elements
'Estes' rough lemon, 150
Ethion, 183
'Etrog' citron, 53
'Eureka' lemon, 46–47
'Eustis' limequat, 18, 61
exocortis, 69, 147–48

fertilizer application: in sprays, 169–71; on the ground, 166–69
fertilizing: bearing trees, 152–71; young trees, 127–30
field capacity of soil, 102
fireguard plowing, 197
firing groves, 211–13
flatwoods soil, 95, 99–100
flavedo, 13
'Florida Common' grapefruit, 34
Florida red scale, 176
Florida scaly bark, 184
foot rot, 67, 68, 69, 75–76, 143–44
Fortunella: as commercial genus, 13, 16; as hybrid parent, 60–61; *crassi-*

folia, 18; *hindsi,* 18; *japonica,* 16–18; *margarita,* 17–18
'Foster' grapefruit, 36
freeze: conditions causing, 84–86; effect of windbreak on, 90; effect on costs and returns, 216; of 1894–95, 6, 85; variable injury by, 85–86
frenching, 161
friendly fungi, 188–89
frost: conditions causing, 84; effect of windbreak on, 90; protection from, by cover, 90
frozen fruit, 214–15
fruit-drop prevention, 191
Fuller rose beetle, 180–81
Fumazone, 151

gibberellic acid (GA), 191
'Glen Improved' naval orange, 28
'Glen Summer' orange, 28
Glover (long) scale, 6, 177
grades for fruit, 227–28
grafting, 79–80
grapefruit: arsenic spray on, 35, 189–91; as stock, 70, 144, 147; canning of, 35–36; mutations of, 33–34; origin of, 31; production of, 32; seedless vs. seedy, 33–34; utilization of, 35–36; varieties of, 36–38
grasshoppers, 180, 196
greasy spot, 182, 186–87
green manure crops, 194
green scale, 178
grove heating, 211–13
grove rehabilitation, 138–42
grove site: planting of, 123–25; preparation, 116–17; spacing, 118–20; staking, 120; system for planting, 117–18; selection: for cold protection, 112–13; for soil drainage, 114; for other factors, 115–16
growth regulators, 191–92

'Hamlin' orange, 26
hammock land, 95
hanging bud, 76, 79
harvesting the crop, 217–19
'Harvey' lemon, 48
heaters for cold protection, 211–12
hedging (pruning), 208–9
hesperidium, 12, 13
high pineland, 95
home plantings of citrus trees: care of, 236–37; for various areas of state, 232; stocks for, 234; transplanting to, 234–36
"honeydew," 178
humidity. *See* relative humidity
hurricanes, 93
hybrids, 14, 53–62

inarching, 82
Indian River citrus area, 86, 107–8
iron as a nutrient, 159–60, 164
irrigation: and cold protection, 213–14; and drainage, 203; equipment, 204; general, 199–204; need of, 202–3; sources of water for, 203; timing of, 202, 203
Isle of Pines grapefruit, 37

'Jaffa' orange, 30
juice content of fruit, 222, 223

Kelthane, 183
'Key' lime, 50, 144
'King' mandarin ("orange"), 39–40
kumquat, 13, 16–18; cold hardiness of, 89; hybrids of, 18, 60–62; meaning of name, 17; origin, 1; uses of, 17; varieties of, 16–18
'Kunenbo', 39
'Kusaie' lime, 53

ladybeetle: Australian (Crypt), 188; blood-red, 188; Chinese, 188; Vedalia, 177, 188
Lakeland fine sand. *See* Astatula fine sand
'Lakeland' limequat ("golden lime"), 18, 61
'Lake' tangelo, 56
lead arsenate spray on grapefruit, 189–91
'Lee' "tangelo," 62, 148
lemon, 'Rough'. *See* 'Rough' lemon
lemon, true: early history of, 45; production of, 45; varieties of, 46–48
leprosis mite, 184
lime: early history of, 50–51; types of, 50–52; world production of, 51–52
limequat, 18, 60–61
limestone (lime), 165
limestone soils, 100
long (Glover) scale, 6, 40, 177
'Lue Gim Gong' orange, 27

macrophylla, 142–43, 147, 148
magnesium as a nutrient, 158–59
malathion, 176, 178
mandarins, 38–44
manganese as a nutrient, 161–62, 164, 171

'Marsh' grapefruit, 36–37
marsh soils, 99–100
maturity of fruit, 219–26; acid content, 224; color break, 221–22; juice content, 222–23; minimum standards for, 220–26; ratio, 224–25; soluble solids, 223–24; spray to hasten, 189–91
meadow (root-lesion) nematode, 151
mealybugs, 178
melanose, 185–86
'Mexican' lime, 50
'Meyer' lemon, 48–49
Microcitrus, 13
'Milam', 150
'Minneola' tangelo, 56, 148, 191
minor element sprays, 169–71
mites: broad, 184–85; citrus bud, 185; citrus red, 183; citrus rust, 182–83; leprosis, 184; purple, 183; six-spotted, 184; Texas citrus, 184
molybdenum as a nutrient, 162, 171
'Morton' citrange, 70
'Murcott' tangor ('Honey Murcott'), 59–60, 148

navel orange, 22, 27–28
Nemagon, 151
nematode: burrowing, 148–51; citrus, 151; root-lesion (meadow), 151
nitrogen as a nutrient, 155–57
'Nova' "tangelo," 62, 148
nucellar embryos, 14
nucellus, 14
nursery for seedlings, 73–74
nursery trees: handling and transporting, 77–78, 121–22; methods of planting, 123–24; selection of, 120–22
nutrient elements: application of, on ground, 166–69; on leaves, 169–71; classes of: macronutrient (major), 153–54; micronutrient (minor), 153–54; primary, 154; secondary, 154; trace, 153; essential: boron, 162; calcium, 159; copper, 160–61; iron, 159–60; magnesium, 158–59; manganese, 161–62; molybdenum, 162; nitrogen, 155–57; phosphorus, 157; potassium, 158; sulfur, 159; zinc, 161; soil reaction and availability of, 163–64
nutritional sprays, 169–71, 189

oil sprays, 175, 177
'Oneco' tangerine, 44
oolitic limestone soil, 100
orange, sour. *See* sour orange
orange, sweet. *See* sweet orange
orange subfamily, 12–14
origin of citrus fruits, 1–4
'Orlando' tangelo, 56, 148, 191
'Osceola' "tangerine," 62, 148, 191
Otaheite-orange, 70
'Owari' satsuma, 39

'Page' "orange," 62, 148
parasitic insects, 187–88
parathion, 175–76, 178, 237
'Parson Brown' orange, 26
'Persian' lime: areas where grown, 51; early history of, 50–51
Phillipe, Dr. Odette, 31
pH of soil, 96, 98–99, 160–62, 163–64
phosphorus as a nutrient, 157
physiological sprays, 189–92
'Pineapple' orange, 25–26, 191
plant bugs, 179–80, 195–96
planting of grove: banking after, 124–25; choice of planting system, 117–18; methods, 123–24; season, 123; site selection, 111; spacing, 118–20; staking tree locations, 120; transplanting from nursery, 120–22
plant-water relations, 201–2
plugging of fruit, 217–18
pomelo, 31, 55
Poncirus: as commercial genus, 13–16; as stock, 69–70; hybrids of, 15–16; *trifoliata*, 15–16
'Ponderosa' lemon, 48
'Ponkan' tangerine, 43–44
'Pope Summer' orange, 28
potash. *See* potassium
potassium as a nutrient, 158
processing, fruit for, 20–21, 229–31
production of grove: basic factors in, 136–37; causes of tree decline and, 142–51; grove rehabilitation and, 138–42; program for, 137
propagation: budding methods for, 75–76; by budding or by seed, 63–64; by inarching, 82; by topworking, 78–82; care of budlings in, 77–78; history in Florida, 63–64; planting seeds for, 72–73; seedlings in nursery for, 72–75; stocks for, 65–71
protection from cold: banking in early grove years for, 134–35; banking newly planted trees for, 124–25; choice of stocks for, 89; economics of, 216; effect of cover on, 90; effect of grove site on, 112; firing in bearing grove for, 211–13; natural, 112; of

budded nursery trees, 77; sprinklers for, 213–14; wind machines for, 214
pruning bearing trees, 204–10; after cold injury, 215–16; deadwood, 205–6; hedging, 208–9; lifting grapefruit, 207; special practices, 206–7; sprouts, 205; techniques of, 210; thinwood, 206; to open up tree top, 206–7; very weak trees, 209–10
pruning young trees, 132
psorosis, 145–46
purple mite, 183
purple scale, 174–76

'Queen' orange, 29

rainfall: distribution in Florida, 91–92; intensity of, 92
'Rangpur' lime, 52–53, 144, 147, 148
ratio of solids to acid, 224–25
red-alga spot, 187
red spider, 183
rehabilitation of grove, 138–42
relative humidity: importance of, 92–93; relation to disease, 93; variation of, in Florida, 92
replacement of grove trees, 141
Ridge area, 96
'Ridge' pineapple, 150
'Robinson' "tangerine," 62, 148, 191
root-lesion (meadow) nematode, 151
root rot, 144–45
rootstocks. *See* stocks
'Rough' lemon, 49–50, 66–67, 142–43, 144
'Royal' grapefruit, 37
'Ruby' blood orange, 30–31
'Ruby' grapefruit, 37
'Rusk' citrange, 16, 61, 70
rust mite, 182–83

'Sampson' tangelo, 57
satsumas, 39, 148; stocks for, 15, 39, 69
scab, 186
scale insects: black, 178; brown soft, 178; chaff, 177; citrus snow, 176; control of, 173–78; cottony-cushion, 177; Florida red, 176; Glover (long), 6, 177; green, 178; purple, 174–76; yellow, 177
seedbed, 72–73
'Seminole' tangelo, 56
shield budding, 75–76
'Sicily' lemon, 47
six-spotted mite, 184
sizes of fruit, 226–27
'Smith' "tangerine," 59
snails for pest control, 189
snow scale, 176
soap for aphid control, 179
soft brown scale, 178
soil amendments, 164–66
soil drainage, 114–15, 202
soil management, 193–98
soil moisture, 97, 98, 102, 200–201
soil reaction, 96, 98, 102–3, 164–66
soils: for citrus growing, 94–103; (classed by drainage: oolitic limestone, 100; poorly drained, 97–99; very poorly drained, 99–100; well-drained, 96–97); classed by native cover, 94–96; desirable depth for rooting of, 114; drainage of, 96, 97, 98, 99, 202; effect of alkalinity of, 98–99; effect of shallowness of, 98–99; exchange capacity of, 97–98; factors in choosing for planting, 100–103; pH of, 96, 98; reaction of, 102–3; water-holding ability of, 102
soil-water relations, 200–201
soluble solids content of fruit, 223–24
sooty mold fungus, 178
sour orange, 2, 5; as stock, 67, 143–44, 146–47
spraying: for control of pests: aphids, 179; citrus root weevil, 181; diseases, 185–87; Fuller rose beetle, 181; grasshoppers, 180, 196; mealybugs, 178; mites, 182–85; plant bugs, 179–80, 195–96; scales, 174–78; sugarcane rootstalk borer weevil, 181; whiteflies, 178–79; for control of fruit drop, 191; for early maturity of grapefruit, 189–91; for minor element deficiency, 169–71, 189; young trees, 130–31
spray program: for bearing grove, 172–73; biological control of pests and, 186–89
spreading decline: cause of, 148–49; control of, 149–51
sprinklers for cold protection, 213–14
sprout pruning, 205
star melanose, 186
'Star Ruby' grapefruit, 34, 38
stink bug, 179–80, 195–96
stocks: properties of desirable, 65; commercial in Florida: 'Carrizo' citrange, 70; 'Cleopatra', 67–68; grapefruit, 70; 'Rough' lemon, 66–67; sour orange, 67; sweet orange, 68–69; trifoliate-orange, 69